STUDIES IN FUNCTIONAL
ANALYSIS

MAA STUDIES IN MATHEMATICS

Published by
THE MATHEMATICAL ASSOCIATION OF AMERICA

———

Committee on Publications
E. F. BECKENBACH, Chairman

Subcommittee on MAA Studies in Mathematics
G. L. WEISS, Chairman
T. M. LIGGETT
A. C. TUCKER

Studies in Mathematics

The Mathematical Association of America

Volume 1: STUDIES IN MODERN ANALYSIS
edited by R. C. Buck

Volume 2: STUDIES IN MODERN ALGEBRA
edited by A. A. Albert

Volume 3: STUDIES IN REAL AND COMPLEX ANALYSIS
edited by I. I. Hirschman, Jr.

Volume 4: STUDIES IN GLOBAL GEOMETRY AND ANALYSIS
edited by S. S. Chern

Volume 5: STUDIES IN MODERN TOPOLOGY
edited by P. J. Hilton

Volume 6: STUDIES IN NUMBER THEORY
edited by W. J. LeVeque

Volume 7: STUDIES IN APPLIED MATHEMATICS
edited by A. H. Taub

Volume 8: STUDIES IN MODEL THEORY
edited by M. D. Morley

Volume 9: STUDIES IN ALGEBRAIC LOGIC
edited by Aubert Daigneault

Volume 10: STUDIES IN OPTIMIZATION
edited by G. B. Dantzig and B. C. Eaves

Volume 11: STUDIES IN GRAPH THEORY, PART I
edited by D. R. Fulkerson

Volume 12: STUDIES IN GRAPH THEORY, PART II
edited by D. R. Fulkerson

Volume 13: STUDIES IN HARMONIC ANALYSIS
edited by J. M. Ash

Volume 14: STUDIES IN ORDINARY DIFFERENTIAL EQUATIONS
edited by Jack Hale

Volume 15: STUDIES IN MATHEMATICAL BIOLOGY, PART I
edited by S. A. Levin

Volume 16: STUDIES IN MATHEMATICAL BIOLOGY, PART II
edited by S. A. Levin

Volume 17: STUDIES IN COMBINATORICS
edited by G.-C. Rota

Volume 18: STUDIES IN PROBABILITY THEORY
edited by Murray Rosenblatt

Volume 19: STUDIES IN STATISTICS
edited by R. V. Hogg

Volume 20: STUDIES IN ALGEBRAIC GEOMETRY
edited by A. Seidenberg

Volume 21: STUDIES IN FUNCTIONAL ANALYSIS
edited by R. G. Bartle

Robert G. Bartle
University of Illinois, Urbana-Champaign

F. F. Bonsall
University of Edinburgh, Scotland

E. W. Cheney
University of Texas, Austin

J. Duncan
University of Stirling, Scotland

William B. Johnson
The Ohio State University, Columbus

R. R. Phelps
University of Washington

H. H. Schaefer
*University of Tübingen
Federal Republic of Germany*

Studies in Mathematics

Volume 21

STUDIES IN FUNCTIONAL ANALYSIS

R. G. Bartle, editor
University of Illinois at Urbana-Champaign

Published and distributed by
The Mathematical Association of America

© *1980 by*
The Mathematical Association of America (Incorporated)
Library of Congress Catalog Card Number 80-81042

Complete Set ISBN 0-88385-100-8
Vol. 21 ISBN 0-88385-121-0

Printed in the United States of America

Current printing (last digit):
10 9 8 7 6 5 4 3 2 1

INTRODUCTION

In the introduction to his article "Preliminaries to functional analysis" that appeared in Volume 1 of these *Studies in Mathematics* (1962), Caspar Goffman wrote: "Functional analysis, briefly, is that branch of mathematics in which elements of a given class of functions are considered to be points of an appropriate infinite dimensional space. In this way, various theories of functions, real and complex, are seen from either a geometric or a point-set theoretic view, often achieving new levels of unification."

While this definition of the term "functional analysis" is by no means perfect, it is probably as good as anyone is likely to formulate. To be sure, the objects of study in functional analysis are not *always* functions (in a natural way, at least) or operators on them; but they *generally* are, and it is this fact that puts functional analysis firmly in the camp of analysis, rather than that of topology. It is true that the mathematicians who, about 1908, laid the foundations of functional analysis, and among whom we count Fréchet, F. Riesz, Hausdorff, and E. Schmidt, were also active in the founding of point-set topology. However, from the very beginning, what we now think of as functional analysis was nourished by the rich analytic work that had been done in a somewhat more classical manner by such giants as Volterra, Fredholm, Hadamard, and Hilbert. F. Riesz, while not the first to formulate axioms closely related to those of a normed linear space, gave a beautifully "modern" discussion of a general form of the Fredholm theory of compact operators in his 1918 *Acta Mathematica* paper. With this paper, "functional analysis" was certainly a reality.

By the end of the 1920's, the work of Lebesgue, Helly, Hahn, Hildebrandt, Banach, and Schauder had developed the main results in the abstract theory of what we now call Banach spaces. Meanwhile, von Neumann and Stone were working on the related Hilbert space theory with an eye toward its application, especially to differential equations. The year 1932 was surely a milestone in the history of functional analysis, for in that year the books of Banach and Stone were both published, and they remained the definitive treatises, at least until the appearance of Hille's "Functional analysis and semi-groups" in 1948. But there had been an explosive development of the field during that time. To mention only some of the highlights, we cite: (i) the development of the general theory of topological vector spaces initiated by Kolmogorov (1934), von Neumann (1935), Dieudonné (1943), Mackey (1946), Arens (1947), et al.; (ii) the fixed point theorems of Schauder (1930), Tihonov (1935), Markov (1936), and Kakutani (1938) and their application to differential equations by Leray and Schauder (1934); (iii) the development of abstract integrals by Bochner (1933), G. Birkhoff (1935), Dunford (1935, 1936), Gel'fand (1938), Pettis (1938), et al.; (iv) the work of Stone (1937) on the representation of Boolean rings, of M. and S. Kreĭn (1940) and Kakutani (1941) on L- and M-spaces, and Gel'fand on "normed rings" (now called "Banach algebras"); (v) the Kreĭn-Mil'man (1940) theorem on extreme points and the theorems of Kreĭn and Šmulian (1939, 1940) on weak compactness and convex sets; (vi) the operational calculus of E. R. Lorch (1942) and Dunford (1943); and (vii) the work of Murray and von Neumann (1936–1943) on rings of operators. Even this list, which is intended to be illustrative rather than exhaustive, does not suggest the wealth of the applications that were beginning to be made of functional analysis in other areas, such as Fourier analysis, harmonic analysis, differential and integral equations, approximation theory, probability theory, ergodic theory, quantum mechanics, and many other areas.

In this small volume we present to the reader five expository accounts of some of the developments that have taken place recently. Recognizing that it is not possible to give a full account (and perhaps not even a balanced one) of the present state of the field in such limited space, we have focused on several rather diverse topics. We believe that the reader will find it fascinating—as we do—to see how varied are the algebraic, analytic, topological, and combinatorial techniques that are employed by the authors of these articles. While these articles are directed somewhat more toward the space than the operators on them, the reader will see a considerable interplay between these notions.

In the first article, Bonsall and Duncan give an account of recent work using the notion of numerical range of operators in Banach spaces. This tool, which at first glance seems too elementary to be promising, is seen—by using some shrewd geometrical and analytic arguments—to have some surprising consequences, among them, a proof of an abstract characterization of C^*-algebras.

In the second article, Cheney gives an overview of certain questions in the theory of approximation. He shows how the desire to settle certain "practical" questions concerning optimal approximation leads to a need to know more about the geometrical nature of certain Banach spaces.

Compact operators have been with functional analysis since the start. Some of the initial studies of compact operators in Hilbert space were based on the fact that such operators can be approximated by finite rank operators, so it has long been a source of embarrassment that the corresponding fact in Banach spaces was not known until Enflo gave a counterexample, in 1973. Johnson presents a description of how such a counterexample can be constructed, using a "problems-with-hints" format. In doing so, he leads the reader into a portion of Banach space structure theory, a field that has been very active recently. While the reader will find some of the problems easy, he will discover that others require considerable combinatorial skill.

The Kreĭn-Mil'man Theorem is one of the major results of functional analysis. Recently, Choquet (1956) proved a result that can be regarded as a refinement of the Kreĭn-Mil'man theorem. In the fourth article, Phelps gives an exposition of Choquet's theorem and its applications. This article is an interesting mixture of geometry and analysis.

While the notion of order was studied in a functional analytic setting by Kantorovič (1937) and others, the theory of ordered spaces has remained somewhat outside of the mainstream of functional analysis until recently. In the final article, Schaefer gives a thorough account of the theory of Banach lattices, including a proof of the Kakutani-Kreĭn representation theorems, and an introduction to tensor products of Banach lattices. He gives a number of examples and applications of these important results.

While there are a number of important areas of functional analysis that are not touched on at all in this study, we feel there is enough variety to interest many readers and that it gives a taste of what has been done recently in functional analysis.

ROBERT G. BARTLE

CONTENTS

INTRODUCTION
Robert G. Bartle, vii

NUMERICAL RANGES
F. F. Bonsall and J. Duncan, 1

PROJECTION OPERATORS IN APPROXIMATION THEORY
E. W. Cheney, 50

COMPLEMENTABLY UNIVERSAL SEPARABLE
BANACH SPACES: AN APPLICATION OF
COUNTEREXAMPLES TO THE APPROXIMATION PROBLEM
William B. Johnson, 81

INTEGRAL REPRESENTATIONS FOR ELEMENTS OF
CONVEX SETS
R. R. Phelps, 115

ASPECTS OF BANACH LATTICES
H. H. Schaefer, 158

INDEX, 223

INDEX OF SYMBOLS, 227

NUMERICAL RANGES

F. F. Bonsall and J. Duncan

1. THE CLASSICAL NUMERICAL RANGE

All mathematicians will have had some contact with quadratic forms; that is, with mappings of the form

$$x \mapsto \langle Tx, x \rangle,$$

where T is an $n \times n$ complex matrix, x is an n-vector, and $\langle \, , \, \rangle$ is the usual inner product

$$\langle x, y \rangle = x_1 y_1^* + \cdots x_n y_n^*,$$

with α^* denoting the complex conjugate of $\alpha \in \mathbb{C}$. They will be familiar with canonical quadratic forms and their application to the study of conics and quadrics. They will have encountered positive quadratic forms and their application to inequalities and max-min problems. But it is likely that they have never met the classical numerical range of T, namely the set $W(T)$ of complex numbers given by

$$W(T) = \{\langle Tx, x \rangle : \langle x, x \rangle = 1\},$$

although it is more than fifty years since O. Toeplitz (1918) and F. Hausdorff (1919) established the interesting fact that $W(T)$ is a convex set.[†]

Although the numerical range remains a useful tool for the study of finite matrices, the main interest in recent years has been in its application to operators on infinite dimensional spaces. For operators T on a complex Hilbert space, the formula for $W(T)$ is immediately applicable. This will be the subject of the present section, and in later sections we consider some generalizations of the concept that are available for operators on Banach spaces.

For the rest of this section H will denote a complex Hilbert space with inner product $\langle\,,\,\rangle$ and norm $\|\ \|$, related by

$$\|x\| = \langle x, x\rangle^{\frac{1}{2}}.$$

The symbol T will denote a bounded linear operator on H, and I will denote the identity operator on H. We recall that there is then a unique bounded linear operator T^*, the adjoint of T, which satisfies $\langle Tx, y\rangle = \langle x, T^*y\rangle$ for all $x, y \in H$. We recall that T is *self-adjoint* if $T^* = T$, is *positive* if it is self-adjoint and $\langle Tx, x\rangle \geq 0$ for all $x \in H$, is *normal* if $T^*T = TT^*$, and is *unitary* if $T^*T = TT^* = I$. The notation $T \geq 0$ denotes that T is positive.

DEFINITION: The *numerical range* of T is the set $W(T)$ of complex numbers given by $W(T) = \{\langle Tx, x\rangle : \langle x, x\rangle = 1\}$. The *numerical radius* $w(T)$ is given by $w(T) = \sup\{|\lambda| : \lambda \in W(T)\}$.

The following properties are evident:

(i) $W(\alpha I + \beta T) = \alpha + \beta W(T)$ $(\alpha, \beta \in \mathbb{C})$;
(ii) $W(T^*) = \{\lambda^* : \lambda \in W(T)\}$;
(iii) $w(T) \leq \|T\|$;
(iv) $|\langle Tx, x\rangle| \leq w(T)\|x\|^2$ $(x \in H)$;
(v) $W(U^*TU) = W(T)$ for all unitary operators U.

[†]Unless otherwise indicated, detailed references to the literature may be found in the authors' books NR I and NR II (see page 48) given in the References.

The first non-trivial property is the Hausdorff-Toeplitz theorem:

(vi) $W(T)$ is a convex set,

which is true for a general complex Hilbert space H. Many elementary proofs of this theorem have been given, and those available by 1971 have been summarized by P. R. Halmos [8].

Many elementary properties of self-adjoint operators follow easily from the following inequality:

(vii) If $T \geqslant 0$, then $\|Tx\|^2 \leqslant w(T)\langle Tx, x \rangle$ for all $x \in H$.

To prove this, consider the bilinear form B given by $B(x, y) = \langle Tx, y \rangle$. The Cauchy-Schwarz inequality applied to this form gives

$$|\langle Tx, y \rangle|^2 \leqslant \langle Tx, x \rangle \langle Ty, y \rangle.$$

Then, taking $y = Tx$ and using (iv), we have

$$\|Tx\|^4 = \langle Tx, Tx \rangle^2 \leqslant \langle Tx, x \rangle \langle T(Tx), Tx \rangle \leqslant \langle Tx, x \rangle w(T) \|Tx\|^2,$$

and (vii) is proved.

It follows at once from (vii) that

$$T \geqslant 0, \quad \langle Tx, x \rangle = 0 \Rightarrow Tx = 0, \tag{1.1}$$

a fact which we shall use several times. In the first place it gives

$$W(T) \subset \mathbb{R} \Leftrightarrow T^* = T. \tag{1.2}$$

The implication $\Leftarrow$ is obvious, since $T^* = T$ gives $\langle Tx, x \rangle = \langle x, Tx \rangle = \langle Tx, x \rangle^*$. To prove $\Rightarrow$, suppose that $W(T) \subset \mathbb{R}$. Then, for all $x \in H$,

$$\langle (T - T^*)x, x \rangle = \langle Tx, x \rangle - \langle x, Tx \rangle = \langle Tx, x \rangle - \langle Tx, x \rangle^* = 0,$$

since $\langle Tx, x \rangle$ is real. It follows that the self-adjoint operator $i(T - T^*)$ is positive and so (1.1) gives $i(T - T^*)x = 0$, whence it follows that $T = T^*$.

As an immediate corollary of (1.2), we have

$$W(T) \subset \mathbb{R}^+ \Leftrightarrow T \geqslant 0, \tag{1.3}$$

and so the condition $T^* = T$ is redundant in the definition of positive operators.

Some of the most important applications of the numerical range concern the *spectrum* of T. We recall that this is the set $\mathrm{Sp}(T)$ of all complex numbers λ such that $\lambda I - T$ does not have an inverse in the algebra of all bounded linear operators on H; it is always a non-void compact set. It contains as a subset (perhaps empty) the *point spectrum* $\mathrm{pSp}(T)$ which is the set of *eigenvalues* of T, in other words those complex numbers λ for which the kernel of $\lambda I - T$ contains non-zero vectors. It is obvious that

$$\mathrm{pSp}(T) \subset W(T); \tag{1.4}$$

for if λ is an eigenvalue of T, there exists a vector $x \in H$ with $\|x\| = 1$ and $Tx = \lambda x$. Then $\lambda = \langle Tx, x \rangle \in W(T)$.

Another subset of the spectrum is the *approximate point spectrum* $\mathrm{apSp}(T)$ which is the set of complex numbers λ such that there exist vectors $x_n \in H$ with $\|x_n\| = 1$ and $\lim_{n \to \infty} \|(\lambda I - T)x_n\| = 0$. Since, for such x_n,

$$|\lambda - \langle Tx_n, x_n \rangle| = |\langle (\lambda I - T)x_n, x_n \rangle| \leq \|(\lambda I - T)x_n\|,$$

we see at once that $\mathrm{apSp}(T) \subset \mathrm{cl}\, W(T)$, where $\mathrm{cl}\, W(T)$ denotes the closure of $W(T)$. However, it is known that the boundary of the spectrum is contained in the approximate point spectrum. Since $W(T)$ is convex, it now follows that

$$\mathrm{co}\, \mathrm{Sp}(T) \subset \mathrm{cl}\, W(T); \tag{1.5}$$

that is, the convex hull of the spectrum is contained in the closure of the numerical range.

EXAMPLE 1: Let H be the two-dimensional space $\mathbb{C}^2$ and let

$$T = \begin{pmatrix} 0 & 1 \\ 0 & 0 \end{pmatrix}.$$

Then

$$W(T) = \{wz^* : z, w \in \mathbb{C}, |z|^2 + |w|^2 = 1\}$$
$$= \{\zeta : \zeta \in \mathbb{C}, |\zeta| \leq \tfrac{1}{2}\},$$

but $\mathrm{Sp}(T) = \{0\}$.

This example shows that equality does not hold, in general, in (1.5). However, for normal operators T we do have

$$\operatorname{co}\operatorname{Sp}(T) = \operatorname{cl} W(T). \tag{1.6}$$

This was proved by A. Wintner [12] in 1929. To prove (1.6) we recall that the *spectral radius* $r(T)$ is given by the formula $r(T) = \lim_{n\to\infty} \|T^n\|^{1/n}$ and that

$$r(T) = \sup\{|\lambda| : \lambda \in \operatorname{Sp}(T)\}. \tag{1.7}$$

It is an elementary property of normal operators that

$$r(T) = \|T\|. \tag{1.8}$$

Let $\Delta(\alpha, \kappa)$ denote the closed disc in the complex plane with center α and radius κ, and suppose that $\operatorname{Sp}(T) \subset \Delta(\alpha, \kappa)$. Then $\operatorname{Sp}(T - \alpha I) \subset \Delta(0, \kappa)$, and so, by (1.7), $r(T - \alpha I) \leq \kappa$. But $T - \alpha I$ is normal. Therefore $\|T - \alpha I\| \leq \kappa$, and (iii) gives $W(T - \alpha I) \subset \Delta(0, \kappa)$, and so $W(T) \subset \Delta(\alpha, \kappa)$. We have proved that every closed disc containing $\operatorname{Sp}(T)$ contains $W(T)$, and therefore $\operatorname{cl} W(T) \subset \operatorname{co}\operatorname{Sp}(T)$. With (1.5), this proves (1.6).

Besides giving a set within which the point spectrum must lie, the numerical range can be used to prove that certain points are eigenvalues. The simplest result of this kind is as follows:

$$\lambda \in W(T), \ |\lambda| = \|T\| \Rightarrow \lambda \in \operatorname{pSp}(T). \tag{1.9}$$

To prove this, we may assume that $\lambda = 1 = \|T\|$. Then there exists $x \in H$ such that $\langle x, x \rangle = 1 = \langle Tx, x \rangle$, and we have

$$2 = \langle Tx + x, x \rangle \leq \|Tx + x\| \leq \|Tx\| + \|x\| \leq 2.$$

Therefore, $\|\tfrac{1}{2}(Tx + x)\| = 1 = \|Tx\| = \|x\|$, and the strict convexity of the unit ball in a Hilbert space implies that $Tx = x$.

A more interesting result of this kind was proved by W. F. Donoghue, Jr. (1957). Given a convex subset K of $\mathbb{C}$, a point $\lambda \in K$ is called a *corner* of K if K is contained in an angle with vertex at λ and magnitude less than π radians.

THEOREM 1: *Let λ be a corner of $\operatorname{cl} W(T)$. Then $\lambda \in \operatorname{Sp}(T)$. If also $\lambda \in W(T)$ then λ is an eigenvalue of T.*

Proof: By (i) we may suppose that $\lambda=0$ and that there exists $\delta>0$ such that

$$\operatorname{Re} W(e^{i\theta}T)\subset\mathbb{R}^+ \qquad (-\delta\leq\theta\leq\delta), \tag{1.10}$$

where $\operatorname{Re} W(e^{i\theta}T)$ denotes the set of real parts of elements of $W(e^{i\theta}T)$. Let $S=T+T^*$. Since $\langle Sx,x\rangle=2\operatorname{Re}\langle Tx,x\rangle$, (1.10) gives $W(S)\subset\mathbb{R}^+$, and so $S\geq 0$. Since $0\in\operatorname{cl} W(T)$, there exist vectors $x_n\in H$ with $\|x_n\|=1$ such that $\lim_{n\to\infty}\langle Tx_n,x_n\rangle=0$. Therefore $\lim_{n\to\infty}\langle Sx_n,x_n\rangle=0$. By (vii), we have

$$\|Sx_n\|^2\leq w(S)\langle Sx_n,x_n\rangle,$$

and so $\lim_{n\to\infty} Sx_n=0$, that is $\lim_{n\to\infty}(T+T^*)x_n=0$. By (1.10), we may replace T by $e^{i\theta}T$ with $-\delta\leq\theta\leq\delta$, and so $\lim_{n\to\infty}(e^{i\theta}T+e^{-i\theta}T^*)x_n=0$ for all such θ. Therefore $\lim_{n\to\infty} Tx_n=0$, and so $0\in\operatorname{Sp}(T)$.

If also $0\in W(T)$ the sequence $\{x_n\}$ is replaced by a vector x with $\|x\|=1$ and $\langle Tx,x\rangle=0$. Then we obtain $Sx=0$ and finally $Tx=0$ as required.

The corners of a convex set K are *extreme points* of K; that is, they are points $\lambda\in K$ such that, whenever $\lambda=\alpha\mu+(1-\alpha)\nu$ with μ, $\nu\in K$ and $0<\alpha<1$, we have $\lambda=\mu=\nu$. We denote by $\operatorname{ext} K$ the set of extreme points of K. Example 1 shows that the extreme points of $W(T)$ need not belong to the spectrum. Nevertheless for certain classes of operators the extreme points of $W(T)$ are eigenvalues. A simple result of this kind concerns self-adjoint operators:

$$T^*=T\Rightarrow\operatorname{ext} W(T)\subset\operatorname{pSp}(T). \tag{1.11}$$

By (i), this is a simple corollary of:

$$T^*=T,\quad 0\in\operatorname{ext} W(T),\quad \langle Tx,x\rangle=0\Rightarrow Tx=0, \tag{1.12}$$

which is a special case of Theorem 1.

A deeper result of this kind was proved by C. H. Meng (1957), namely, that the conclusion (1.11) holds for normal operators. Then J. G. Stampfli [11] obtained the same conclusion for the wider class of hyponormal operators; T is said to be *hyponormal* if $T^*T-TT^*\geq 0$.

THEOREM 2: *Let T be hyponormal. Then* $\text{ext } W(T) \subset \text{pSp}(T)$.

Proof: The following proof of this theorem is a variant of Stampfli's proof, in which we show that the theorem follows from (1.12). We prove first that for all operators T, we have

$$\text{Im } W(T) \subset \mathbb{R}^+, \langle Tx, x \rangle = 0 \Rightarrow T^*x = Tx, \quad (1.13)$$

where $\text{Im } W(T)$ denotes the set of imaginary parts of the elements of $W(T)$. For all $y \in H$, we have

$$\left\langle \frac{1}{i}(T - T^*)y, y \right\rangle = \frac{1}{i}(\langle Ty, y \rangle - \langle y, Ty \rangle) = 2\text{Im}\langle Ty, y \rangle.$$

Given that $\text{Im } W(T) \subset \mathbb{R}^+$ and $\langle Tx, x \rangle = 0$, this shows that

$$\frac{1}{i}(T - T^*) \geq 0 \quad \text{and that} \quad \left\langle \frac{1}{i}(T - T^*)x, x \right\rangle = 0.$$

Thus (1.1) gives $(1/i)(T - T^*)x = 0$, and (1.13) is proved.

Suppose now that T is hyponormal. Then

$$Tx = T^*x \Rightarrow TT^*x = T^*Tx. \quad (1.14)$$

Indeed, $\langle (T^*T - TT^*)x, x \rangle = \langle Tx, Tx \rangle - \langle T^*x, T^*x \rangle = 0$. But $T^*T - TT^* \geq 0$, since T is hyponormal, and so (1.1) implies (1.14).

Let $N = \{x \in H : T^*x = Tx\}$; we claim that N is a reducing subspace for T and the restriction $T|N$ is self-adjoint. For let $x \in N$; then $Tx = T^*x$ and so (1.14) gives $TT^*x = T^*Tx$. But also $T(T^*x) = T(Tx)$ and $T^*(T^*x) = T^*(Tx)$, since $T^*x = Tx$. Therefore $T^*(Tx) = T(Tx)$, which shows that $Tx \in N$; and $T^*(T^*x) = T(T^*x)$, giving $T^*x \in N$. Thus N is an invariant subspace for T and for T^*; in other words, it is a reducing subspace for T. It is now clear from the definition of N that $T|N$ is self-adjoint.

It is straightforward to verify that $\alpha I + \beta T$ is hyponormal for all complex numbers α, β. Therefore we may assume that $0 \in \text{ext } W(T)$, that $\text{Im } W(T) \subset \mathbb{R}^+$ and that $\langle Tx, x \rangle = 0$ for some vector x with $\|x\| = 1$, and complete the proof by showing that $Tx = 0$. By (1.13), $x \in N$. Let $S = T|N$; then S is self-adjoint, $0 \in \text{ext } W(S)$ (since $W(S) \subset W(T)$) and $\langle Sx, x \rangle = 0$. Thus by (1.12) $Sx = 0$, that is, $Tx = 0$, and the proof is complete.

We have already noted in (v) that $W(T)$ is a unitary invariant. It is far from being a similarity invariant. Indeed, S. Hildebrandt (1966) proved that for all operators T,

$$\operatorname{co} \operatorname{Sp}(T) = \bigcap \{\operatorname{cl} W(S^{-1}TS) : S \text{ invertible}\}.$$

J. P. Williams (1969) strengthened this result by proving that for each open convex set Z containing $\operatorname{Sp}(T)$ there exists an invertible operator S for which $W(S^{-1}TS) \subset Z$. This is a far-reaching generalization of Lyapunov's characterization of $n \times n$ matrices, all of whose eigenvalues λ satisfy $\operatorname{Re}\lambda > 0$.

So far we have been concerned with relations between the spectrum and the numerical range. There are also relations between the norm and the numerical radius. For example:

$$T^* = T \Rightarrow \|T\| = w(T). \tag{1.15}$$

This follows easily from the fact that $r(T) = \|T\|$ for self-adjoint operators T. Alternatively, by (iii) it is enough to prove that $\|T\| \leq w(T)$. Suppose that $w(T) \leq 1$; then $\pm \langle Tx, x \rangle \leq \langle x, x \rangle$, and so $I \pm T \geq 0$. By (vii) we have

$$\|(I \pm T)x\|^2 \leq 2\langle (I \pm T)x, x \rangle,$$

and so the parallelogram rule gives

$$2\|x\|^2 + 2\|Tx\|^2 = \|x + Tx\|^2 + \|x - Tx\|^2 \leq 4\langle x, x \rangle = 4\|x\|^2,$$

whence it follows that $\|Tx\| \leq \|x\|$ and $\|T\| \leq 1$.

By taking real and imaginary parts, we easily deduce from (1.15) that, for all operators T,

$$\|T\| \leq 2w(T), \tag{1.16}$$

and this can also be proved directly by applying the polarization formula to $\langle Tx, y \rangle$. Example 1 shows that the constant 2 here is best possible.

Inequality (1.16) shows that

$$w(T) = 0 \Rightarrow T = 0.$$

In fact, the numerical radius is a linear norm on the space of all bounded operators, the inequality $w(S+T) \leq w(S)+w(T)$ being a simple consequence of the inclusion $W(S+T) \subset W(S)+W(T)$. On the other hand, the inequality $w(ST) \leq w(S)w(T)$ is, in general, false even if $ST = TS$. It is not even true that $w(T^{n+m}) \leq w(T^n)w(T^m)$, as can be proved by considering the matrix

$$\begin{bmatrix} 0 & 1 & 0 & 0 \\ 0 & 0 & 1 & 0 \\ 0 & 0 & 0 & 1 \\ 0 & 0 & 0 & 0 \end{bmatrix}.$$

Consequently, there was some surprise when C. Berger (see [3]) proved the following 'power inequality.'

THEOREM 3: $w(T^n) \leq w(T)^n$ for $n = 1, 2, 3, \ldots$.

The following ingenious and elementary proof of this theorem is due to C. Pearcy (1966). We assume, without real loss of generality, that $w(T) \leq 1$ and prove that $w(T^n) \leq 1$. Note that

$$w(S) \leq 1 \Leftrightarrow \operatorname{Re}\langle (I - e^{i\theta}S)x, x \rangle \geq 0 \quad \text{for } x \in H, \theta \in \mathbb{R}. \quad (1.17)$$

Let $\alpha = \exp(2\pi i/n)$ and let p_k be the complex polynomial given by

$$p_k(z) = 1 + \alpha^k z + \alpha^{2k} z^2 + \cdots + \alpha^{(n-1)k} z^{n-1}.$$

Then we have the identities

$$(1 - \alpha^k z) p_k(z) = 1 - z^n, \quad \sum_{k=1}^{n} p_k(z) = n. \quad (1.18)$$

Since $w(T) \leq 1$, (1.17) gives

$$\operatorname{Re}\langle (I - \alpha^k T) p_k(T) x, p_k(T) x \rangle \geq 0$$

for $x \in H$ and $k = 1, \ldots, n$; and then (1.18) gives

$$\operatorname{Re}\langle (I - T^n) x, x \rangle \geq 0$$

for $x \in H$. Now replace T by $e^{i\theta/n}T$ and apply (1.17).

2. NUMERICAL RANGES OF OPERATORS ON NORMED LINEAR SPACES

The numerical range of an operator on a Hilbert space is defined in terms of the inner product $\langle\ ,\ \rangle$, and so it is not at all evident how the concept should be defined on a normed space where no inner product is available. We shall see that many different numerical ranges can be defined which can be regarded as a bag of tools to be selected for different purposes.

In this section, the symbol X with norm $\|\ \|$ will denote a non-zero complex normed linear space, $S(X)$ its unit sphere $\{x \in X : \|x\| = 1\}$, X' its dual space of all continuous linear functionals on X, and $BL(X)$ the space of all bounded linear operators on X. Our aim is to assign a set of complex numbers to each $T \in BL(X)$, and so it is natural to make use of linear functionals on X and to define the numerical range of T as a set of complex numbers of the form

$$f(Tx)$$

with $x \in X$ and $f \in X'$ chosen in some way. If we examine the classical numerical range from this point of view some possibilities soon become apparent.

Given a Hilbert space H and a vector $y \in H$, let f_y denote the linear functional on H defined for $x \in H$ by

$$f_y(x) = \langle x, y \rangle.$$

We have $f_y(y) = \|y\|^2$ and $|f_y(x)| \leq \|x\| \|y\|$, and so f_y is a continuous linear functional on H with $\|f_y\| = \|y\|$. For an operator T on H, the numerical range is given by

$$W(T) = \{f_y(Ty) : y \in S(H)\}.$$

When $y \in S(H)$, we have $\|y\| = 1$, and so $f_y(y) = 1 = \|f_y\|$, and, moreover, f_y is the only linear functional $f \in H'$ such that $f(y) = 1 = \|f\|$. Thus

$$W(T) = \{f(Tx) : x \in S(H), f \in S(H'), f(x) = 1\}.$$

In this form the concept is immediately available for general normed spaces.

DEFINITION: Let $T \in BL(X)$. The *spatial numerical range* $V(T)$ is defined by

$$V(T) = \{f(Tx) : x \in S(X), f \in S(X'), f(x) = 1\}.$$

Given $x \in S(X)$, let $D(x) = \{f \in S(X') : f(x) = 1\}$. Then $V(T)$ takes the form

$$V(T) = \{f(Tx) : f \in D(x), x \in S(X)\}.$$

When X is a Hilbert space, we have seen that $D(x)$ has exactly one element; when X is a normed space, the Hahn-Banach theorem gives the existence of at least one element in $D(x)$ but, in general, there may be more than one.

An alternative definition of the numerical range, given by G. Lumer (1961), seems at first to be very different. It depends on generalizing the notion of inner product as follows.

DEFINITION: A *semi-inner product* on X is a mapping [,] of $X \times X$ into $\mathbb{C}$ such that:

(i) the mapping $x \mapsto [x, y]$ is linear on X for each fixed $y \in X$;
(ii) $[x, x] > 0$ when $x \neq 0$;
(iii) $|[x, y]|^2 \leq [x, x][y, y]$ for all $x, y \in X$.

A semi-inner product differs from an inner product in lacking the symmetry property $\langle y, x \rangle = \langle x, y \rangle^*$ and, correspondingly, it is not in general antilinear in the second variable. For inner products the axiom (iii) is a theorem (the Cauchy-Schwarz inequality). Given a semi-inner product [,] on X, it can be proved that the mapping $x \mapsto [x, x]^{1/2}$ is a norm on X. If this norm agrees with the given norm on X, we say that the semi-inner product *determines the norm on* X. With such a semi-inner product, the axiom (iii) becomes

$$|[x, y]| \leq \|x\| \|y\| \quad \text{for } x, y \in X,$$

and the numerical range is defined just as for Hilbert spaces.

DEFINITION: Let [,] be a semi-inner product that determines the norm of X, then the corresponding numerical range $W(T)$ of $T \in BL(X)$ is given by $W(T) = \{[Tx, x] : x \in S(X)\}$.

It is not immediately obvious that there exists a semi-inner product that determines the norm of X. However, since $D(x)$ is a non-void set for each $x \in S(X)$, the axiom of choice gives the existence of a mapping $\psi : S(X) \to S(X')$ such that

$$\psi(x) \in D(x) \quad \text{for } x \in S(X). \tag{2.1}$$

Each such mapping ψ gives us a semi-inner product of the right kind, if we construct $[\ ,\ ]$ on $X \times X$ by taking, for all $x \in X$:

$$[x, y] = 0 \quad \text{for } y = 0,$$

$$[x, y] = \|y\| \psi(\|y\|^{-1} y)(x) \quad \text{for } y \neq 0.$$

It is useful to observe that the numerical range $W(T)$ corresponding to this semi-inner product is given by

$$W(T) = \{\psi(x)(Tx) : x \in S(X)\}. \tag{2.2}$$

Conversely, if $[\ ,\ ]$ is any semi-inner product that determines the norm of X, then a mapping ψ of $S(X)$ into $S(X')$ satisfying (2.1) is given by taking

$$\psi(y)(x) = [x, y] \quad \text{for } y \in S(X) \quad \text{and} \quad x \in X.$$

Since $\psi(x) \in D(x)$, it is immediate from (2.2) that

$$W(T) \subset V(T).$$

The following example shows that $W(T)$ may be quite a thin subset of $V(T)$.

EXAMPLE 2: Let $X = C[0, 1]$, the Banach space of continuous complex functions on the unit interval $[0, 1]$ normed with the uniform norm $\|\ \|_\infty$, defined for $x \in X$ by

$$\|x\|_\infty = \sup\{|x(t)| : 0 \leq t \leq 1\}.$$

For each $x \in S(X)$ choose a point $t_x \in [0, 1]$ at which the function $x(\cdot)$ attains its norm, i.e., for which $|x(t_x)| = 1$. Then let $\psi(x)$

be a scalar multiple of the evaluation functional at t_x as follows:

$$\psi(x)(y) = x(t_x)^* y(t_x) \quad \text{for } y \in X.$$

It is easy to check that $\psi(x) \in D(x)$ and so ψ is a mapping of $S(X)$ into $S(X')$ satisfying (2.1). Thus the numerical range of $T \in BL(X)$ for the corresponding semi-inner product is given by

$$W(T) = \{x(t_x)^*(Tx)(t_x) : x \in S(X)\}.$$

In particular, let $\phi \in C[0,1]$ and let M_ϕ denote the multiplication operator on X given by

$$(M_\phi x)(t) = \phi(t) x(t) \quad (t \in [0,1], x \in X).$$

Then we have

$$W(M_\phi) = \{\phi(t_x) : x \in S(X)\}.$$

(a) Given $t \in [0,1]$, we can choose a function $x \in S(X)$ that attains its norm only at t, and so has $t_x = t$. Thus

$$W(M_\phi) = \{\phi(t) : t \in [0,1]\};$$

hence, this numerical range of the operator M_ϕ is the curve given by the range of ϕ.

(b) On the other hand, if μ is any probability measure on $[0,1]$, then the corresponding functional on X is an element of $D(1)$ where 1 is the function with the constant value 1. Thus

$$\int_0^1 \phi(t)\,d\mu(t) = \mu(\phi) = \mu(M_\phi 1) \in V(M_\phi);$$

and so $V(M_\phi)$ contains the convex hull of the range of ϕ.

Choosing a semi-inner product so that the numerical range $W(T)$ is thin can be useful in applications; for example, in locating the eigenvalues of an operator. On the other hand, the arbitrary choice of an element of $D(x)$ that is involved in defining a semi-inner product in general produces a numerical range with less structure than is present in $V(T)$.

Linear isometries on a Hilbert space H are characterized by the condition

$$\langle Tx, Ty \rangle = \langle x, y \rangle \qquad (x, y \in H).$$

A similar result for normed spaces was proved by D. O. Koehler and P. Rosenthal (1970). A linear operator $T: X \to X$ is an isometry if and only if there exists a semi-inner product $[\,,\,]$ determining the norm of X such that

$$[Tx, Ty] = [x, y] \qquad (x, y \in X). \tag{2.3}$$

That T satisfying (2.3) is an isometry is clear, for we have

$$\|Tx\|^2 = [Tx, Tx] = [x, x] = \|x\|^2.$$

Suppose on the other hand that T is an isometry, and let $\langle\,,\,\rangle$ be any semi-inner product that determines the norm of X. Then

$$|\langle T^n x, T^n y \rangle| \leq \|T^n x\| \, \|T^n y\| = \|x\| \, \|y\|,$$

and so the sequence $\{\langle T^n x, T^n y \rangle\}$ is a bounded sequence of complex numbers. By a theorem of S. Banach there exists a *Banach limit* on the space l_∞ of all bounded complex numbers; that is, a linear functional Ψ on l_∞ of norm 1 which agrees with the usual limit for convergent sequences and which is translation invariant in the sense that $\Psi(\{\xi_{n+1}\}) = \Psi(\{\xi_n\})$ for every sequence $\{\xi_n\} \in l_\infty$. Let Ψ be such a Banach limit and define $[\,,\,]$ by

$$[x, y] = \Psi(\{\langle T^n x, T^n y \rangle\}).$$

Since Ψ has norm 1, we have

$$|[x, y]| \leq \sup_n |\langle T^n x, T^n y \rangle| \leq \|x\| \, \|y\|.$$

Also $\langle T^n x, T^n x \rangle = \|T^n x\|^2 = \|x\|^2$, and so $\Psi(\{\langle T^n x, T^n x \rangle\}) = \|x\|^2$. It is now easy to see that $[\,,\,]$ is a semi-inner product determining the norm of X, and the translation invariance of Ψ shows that it satisfies (2.3).

3. PROPERTIES OF NUMERICAL RANGES ON NORMED SPACES

We have seen that an operator on a normed space has many numerical ranges. To develop the properties that these numerical ranges have in common, it is convenient to make use of the axiomatic method. We therefore consider a mapping $T \mapsto \Phi(T)$ defined on $BL(X)$, where, for each T, $\Phi(T)$ is a set of complex numbers satisfying certain axioms.

We use the same notation as in §2 and also, given a set E of complex numbers, we denote by $|E|$ and $\operatorname{Re} E$ the set of moduli and real parts (respectively) of the elements of E. We denote the identity operator on X by I.

DEFINITION: A mapping Φ defined on $BL(X)$ is a *printer* on $BL(X)$ if, for each $T \in BL(X)$, it satisfies the following axioms:

(i) $\Phi(T)$ is a non-void set of complex numbers;
(ii) $\Phi(\alpha I + \beta T) = \alpha + \beta \Phi(T)$ for $\alpha, \beta \in \mathbb{C}$;
(iii) $\sup |\Phi(T)| \leq \|T\|$;
(iv) $\inf |\Phi(T)| \leq \|Tx\|$ for $x \in S(X)$.

If [,] is a semi-inner product that determines the norm of X, the corresponding numerical range W gives a printer $T \mapsto W(T)$ on $BL(X)$. The axioms (i), (ii), (iii) are easily checked; for example (iii) comes from the inequality

$$|[Tx,x]| \leq \|Tx\| \|x\| \leq \|T\| \qquad (x \in S(X)). \tag{3.1}$$

To prove (iv), let $x \in S(X)$. Then $[Tx,x] \in W(T)$, and the first inequality in (3.1) gives $\|[Tx,x]\| \leq \|Tx\|$. Thus (iv) holds. A similar argument shows that $T \mapsto V(T)$ is a printer on $BL(X)$.

Another example of a printer is given by the algebra numerical range which we now define.

DEFINITION: Let B denote the normed linear space $BL(X)$. Then the *algebra numerical range* $V(B,T)$ is defined, for each $T \in B$, by

$$V(B,T) = \{\phi(T) : \phi \in D(I)\}.$$

Here, as in §2, $D(I)$ denote the set of linear functionals on B given by

$$D(I) = \{\phi \in B' : \|\phi\| = 1 = \phi(I)\}.$$

That $T \mapsto V(B, T)$ satisfies (i), (ii), (iii) is clear; to prove (iv) it is enough to prove that

$$V(T) \subset V(B, T). \qquad (3.2)$$

To see this, let $x \in S(X)$ and $f \in D(x)$, and define ϕ on B by $\phi(T) = f(Tx)$. Then $|\phi(T)| \leq \|f\| \|Tx\| \leq \|T\|$ and $\phi(I) = f(x) = 1$. Thus $\phi \in D(I)$ and $f(Tx) \in V(B, T)$. Thus (3.2) holds, and the mapping $T \mapsto V(B, T)$ is a printer.

For the rest of this section we assume that Φ is a printer on $BL(X)$ and that $T \in BL(X)$. With λ a complex number, we denote by $d(\lambda, \Phi(T))$ the distance of λ from $\Phi(T)$. Many results follow easily from the simple observation that

$$\|(\lambda I - T)x\| \geq d(\lambda, \Phi(T)) \|x\| \qquad (x \in X). \qquad (3.3)$$

By homogeneity, it is enough to prove this when $x \in S(X)$. But then (iv) applied to the operator $\lambda I - T$ gives

$$\|(\lambda I - T)x\| \geq \inf |\Phi(\lambda I - T)|$$
$$= \inf |\lambda - \Phi(T)| = d(\lambda, \Phi(T)).$$

It follows at once from the inequality (3.3) that the approximate point spectrum satisfies

$$\mathrm{apSp}(T) \subset \mathrm{cl}\, \Phi(T); \qquad (3.4)$$

for when $\lambda \in \mathrm{apSp}(T)$, there exist $x_n \in S(X)$ such that we have $\lim_{n \to \infty} \|(\lambda I - T)x_n\| = 0$, and this with (3.3) gives $d(\lambda, \Phi(T)) = 0$.

When X is a Banach space, all the boundary points of the spectrum belong to the approximate point spectrum, and so (3.4) gives

$$\partial \mathrm{Sp}(T) \subset \mathrm{cl}\, \Phi(T). \qquad (3.5)$$

The point spectrum is obviously contained in the approximate point spectrum, and so (3.4) gives

$$\text{pSp}(T) \subset \text{cl}\,\Phi(T).$$

All the examples of printers that we have discussed above have the stronger property

$$\text{pSp}(T) \subset \Phi(T). \tag{3.6}$$

This is because these printers satisfy the stronger axiom:

(iv′) For each $x \in S(X)$ there exists $\zeta \in \Phi(T)$ with $|\zeta| \leq \|Tx\|$.

Given $\lambda \in \text{pSp}(T)$, there exists $x \in S(X)$ with $(\lambda I - T)x = 0$. Then (iv′) gives $0 \in \Phi(\lambda I - T)$, $\lambda \in \Phi(T)$. This is as far as we can go in relating the spectrum to a general printer, but we shall see later that much more can be proved for the spatial numerical range $V(T)$.

The following lemma gives a very useful explicit formula for the supremum of the real parts of the elements of $\Phi(T)$ in terms of the operator norm.

LEMMA 1: $\sup \text{Re}\,\Phi(T) = \inf_{\alpha > 0} \alpha^{-1}(\|I + \alpha T\| - 1)$
$\qquad\qquad\qquad = \lim_{\alpha \to 0+} \alpha^{-1}(\|I + \alpha T\| - 1).$

Proof: Let $\mu = \sup \text{Re}\,\Phi(T)$ and $\alpha > 0$. If $\lambda \in \Phi(T)$, we have $\lambda \in \alpha^{-1}\{\Phi(I + \alpha T) - 1\}$, and so

$$\text{Re}\,\lambda \leq \alpha^{-1}\{\sup \text{Re}\,\Phi(I + \alpha T) - 1\}$$
$$\leq \alpha^{-1}\{\sup|\Phi(I + \alpha T)| - 1\} \leq \alpha^{-1}(\|I + \alpha T\| - 1);$$

which gives

$$\mu \leq \inf_{\alpha > 0} \alpha^{-1}(\|I + \alpha T\| - 1). \tag{3.7}$$

Suppose now $\alpha > 0$ and $\alpha^{-1} > \mu$. Then $d(\alpha^{-1}, \Phi(T)) \geq \alpha^{-1} - \mu$, and so (3.3) gives

$$\|(\alpha^{-1}I - T)x\| \geq (\alpha^{-1} - \mu)\|x\| \qquad (x \in X);$$

that is,
$$\|(I-\alpha T)x\| \geq (1-\alpha\mu)\|x\| \qquad (x\in X). \tag{3.8}$$

Replacing x by $(I+\alpha T)x$, we now have
$$\|(I+\alpha T)x\| \leq (1-\alpha\mu)^{-1}\|(I-\alpha^2 T^2)x\| \tag{3.9}$$

for $\alpha > 0$, $\alpha^{-1} > \mu$. It follows that, for all sufficiently small $\alpha > 0$, we have
$$\|I+\alpha T\| \leq (1-\alpha\mu)^{-1}(1+\alpha^2\|T^2\|),$$
$$\alpha^{-1}(\|I+\alpha T\|-1) \leq (1-\alpha\mu)^{-1}(\mu+\alpha\|T^2\|),$$

and so
$$\limsup_{\alpha\to 0+} \alpha^{-1}(\|I+\alpha T\|-1) \leq \mu.$$

With (3.7), this completes the proof of the lemma.

Since $\Phi(\alpha I + \beta T) = \alpha + \beta\Phi(T)$, the lemma shows that the class of closed half planes containing $\Phi(T)$ is independent of the choice of the printer Φ, and so $\text{cl co}\,\Phi(T)$, the closed convex hull of $\Phi(T)$, is also independent of the choice of Φ. The following theorem identifies this closed convex hull.

THEOREM 4: *For all printers Φ on $BL(X)$, $\text{cl co}\,\Phi(T) = V(B,T)$.*

Proof: The set $D(I)$ can be written as an intersection
$$D(I) = \{\phi \in B' : \|\phi\| \leq 1\} \cap \{\phi \in B' : \phi(I) = 1\}.$$

The first of the sets in this intersection is convex and compact in the weak* topology, and the other set is convex and closed in the weak* topology. Therefore $D(I)$ is convex and weak* compact. The mapping $\phi \mapsto \phi(T)$ is linear and weak* continuous, and maps $D(I)$ onto $V(B,T)$. Therefore $V(B,T)$ is convex and compact. But $T \mapsto V(B,T)$ is a printer, and we have seen that all printers Φ give the same set $\text{cl co}\,\Phi(T)$.

NUMERICAL RANGES

The following theorem gives another useful formula for the supremum of the real parts of the elements of $\Phi(T)$.

THEOREM 5: *Let X be a Banach space. Then*

$$\sup \operatorname{Re} \Phi(T) = \sup_{\alpha > 0} \alpha^{-1} \log \|\exp(\alpha T)\| = \lim_{\alpha \to 0+} \alpha^{-1} \log \|\exp(\alpha T)\|.$$

Proof: Let $\mu = \sup \operatorname{Re} \Phi(T)$. With $\alpha > 0$ and $\alpha^{-1} > \mu$, iteration of (3.8) gives

$$\|(I - \alpha T)^n x\| \geq (1 - \alpha \mu)^n \|x\|.$$

Given $\alpha > 0$, we have $(\alpha/n)^{-1} > \mu$ for all sufficiently large n, and so

$$\left\|\left(I - \frac{\alpha}{n} T\right)^n x\right\| \geq \left(1 - \frac{\alpha}{n} \mu\right)^n \|x\|.$$

Letting $n \to \infty$, we obtain $\|\exp(-\alpha T) x\| \geq \exp(-\alpha \mu) \|x\|$, and so

$$\|\exp(\alpha T)\| \leq \exp(\alpha \mu) \quad \text{for } \alpha > 0. \tag{3.10}$$

On the other hand, we have $\|\exp(\alpha T)\| = \|I + \alpha T\| + \lambda(\alpha)$, where, for some positive constant M,

$$|\lambda(\alpha)| \leq M\alpha^2 \quad \text{for } 0 \leq \alpha \leq 1. \tag{3.11}$$

Since $\log t \geq t^{-1}(t - 1)$ for $t > 0$, we have, for $\alpha > 0$,

$$\alpha^{-1} \log \|\exp(\alpha T)\| \geq \frac{\alpha^{-1}\{\|I + \alpha T\| - 1\} + \alpha^{-1}\lambda(\alpha)}{\|I + \alpha T\| + \lambda(\alpha)}.$$

By Lemma 1 and (3.11), the right-hand side of this inequality converges to μ as $\alpha \to 0+$. With (3.10) this completes the proof of the theorem.

We have seen that an operator T on a Hilbert space is self-adjoint if and only if its numerical range is real. Thus numerical ranges make it possible to generalize this important class of operators to general normed spaces, as in the following definition.

DEFINITION: An operator T is *Hermitian* if $\Phi(T)\subset\mathbb{R}$.

On the face of it the set of Hermitian operators depends on the choice of the printer Φ, but since $\mathrm{cl\,co}\,\Phi(T)$ is the same for all printers Φ, we have $\Phi(T)\subset\mathbb{R}$ for all printers Φ if this holds for one printer. Thus the set of Hermitian operators is independent of the choice of printer. The following corollary determines this set directly in terms of the operator norm.

COROLLARY: *Let X be a Banach space. Then T is Hermitian if and only if $\|\exp(itT)\|=1$ for all $t\in\mathbb{R}$.*

Proof: Apply Theorem 5 with T replaced by $\pm iT$.

We shall return to the subject of Hermitian operators in §5, where we establish some of their interesting properties.

Besides locating the eigenvalues of operators, numerical ranges can give information about their structure, as in the following theorem of Nirschl and Schneider (1964), which asserts that every eigenvalue of an operator T belonging to the boundary of $\mathrm{cl\,co}\,\Phi(T)$ has *ascent* 1.

THEOREM 6: *Let λ be a boundary point of $\mathrm{cl\,co}\,\Phi(T)$ and let $(\lambda I - T)^2 x = 0$. Then $(\lambda I - T)x = 0$.*

Proof: Since $\Phi(\alpha I+\beta T)=\alpha+\beta\Phi(T)$, we may suppose that $\lambda=0=\sup\mathrm{Re}\,\Phi(T)$. Then, since $T^2 x=0$, inequality (3.9) implies that

$$\|(I+\alpha T)x\|\leq\|x\|\qquad(\alpha>0),$$

and therefore

$$\alpha\|Tx\|\leq 2\|x\|\qquad(\alpha>0),$$

which gives $\|Tx\|=0$, as required.

The same conclusion holds for boundary points λ of $V(T)$. This is a deeper result due to M. J. Crabb (1970).

If x, y are orthogonal vectors in a Hilbert space, we have $\|x+y\|^2 = \|x\|^2 + \|y\|^2$, and so

$$\|x+y\| \geq \|x\|.$$

This property can be used to define a concept of orthogonality in a normed space. Given linear subspaces E, F of X, we write $E \perp F$ to denote that

$$\|x+y\| \geq \|x\| \quad (x \in E, y \in F).$$

Note, however, that this notion of orthogonality is not a symmetrical one; that is, $E \perp F$ does not imply $F \perp E$.

THEOREM 7 (Crabb and Sinclair (1972)): *Suppose that 0 is not an interior point of* $\operatorname{cl}\operatorname{co}\Phi(T)$. *Then*

$$\|x + Ty\| \geq \|x\| - \sqrt{8\|Tx\|\|y\|} \quad \text{for all } x, y \in X. \quad (3.12)$$

Proof: Since multiplication of T by a complex number of modulus 1 does not change the inequality to be proved, we may suppose that $\Phi(T)$ is contained in the closed left half plane $\{\lambda \in \mathbb{C} : \operatorname{Re}\lambda \leq 0\}$. Then, for all $\alpha \geq 0$, we have $d(\alpha, \Phi(T)) \geq \alpha$, and so (3.3) gives

$$\|(T - \alpha I)z\| \geq \alpha\|z\| \quad (z \in X).$$

The required inequality is obvious if $\|x + Ty\| \geq \|x\|$ or if $y = 0$. Suppose then that $x, y \in X$ with $\|x + Ty\| < \|x\|$ and $y \neq 0$. Taking $z = x - \alpha y$, we obtain

$$\|\alpha^2 y - \alpha(x + Ty) + Tx\| \geq \alpha\|x - \alpha y\|,$$

which gives

$$\alpha^2\|y\| + \alpha\|x + Ty\| + \|Tx\| \geq \alpha\|x\| - \alpha^2\|y\|.$$

Thus, for all $\alpha \geq 0$, we have

$$A\alpha^2 - B\alpha + C \geq 0,$$

where $A = 2\|y\|$, $B = \|x\| - \|x + Ty\|$, and $C = \|Tx\|$. We are assuming that $B > 0$, and so we may take $\alpha = B/2A$ and obtain $4AC - B^2 \geq 0$. This gives $\sqrt{8\|Tx\|\|y\|} = \sqrt{4AC} \geq B$, as required.

COROLLARY: *Suppose that 0 is not an interior point of* $\operatorname{cl} \operatorname{co} \Phi(T)$. *Then* $\ker T \perp TX$.

Proof: If $x \in \ker T$ and $y \in TX$, we have $Tx = 0$ and $y = Tz$ for some $z \in X$. Therefore, by Theorem 7,

$$\|x + y\| = \|x + Tz\| \geq \|x\|.$$

Inequality (3.12) and the corollary that $\ker T \perp TX$ have also been proved by Crabb and Sinclair under the weaker hypothesis that 0 is not an interior point of $V(T)$. This is harder to prove; a full account is given in NR II, §20. The proof of Crabb and Sinclair applies equally to the situation in which T is an unbounded operator defined on all of a Banach space X. Crabb [6] has noted the corollary that $V(T) = \mathbb{C}$ for such an operator. To see this, choose $\{x_n\}$ such that $\lim_{n \to \infty} x_n = 0$, $\lim_{n \to \infty} Tx_n = -x$, and $x \neq 0$. Then (3.12) with $y = x_n$ fails for n sufficiently large and hence $0 \in V(T)$. Since $T - \lambda I$ is unbounded for any $\lambda \in \mathbb{C}$, it follows that $0 \in V(T - \lambda I)$, $\lambda \in V(T)$.

We end this section with some applications of numerical ranges to matrix problems.

EXAMPLE 3: The *Gershgorin discs* of an $n \times n$ complex matrix (a_{ij}) are the closed discs $G_1, \ldots, G_n$ in the complex plane given by

$$G_k = \{z \in \mathbb{C} : |z - a_{kk}| \leq r_k\},$$

where $r_k = \sum_{j=1}^n |a_{kj}| - |a_{kk}|$. We prove that *the eigenvalues of a matrix are contained in the union of its Gershgorin discs.*

Let $X = \mathbb{C}^n$ with the norm $\| \ \|_\infty$ given by $\|\mathbf{x}\|_\infty = \max\{|x_k| : 1 \leq k \leq n\}$, where $\mathbf{x} = (x_1, x_2, \ldots, x_n) \in X$, and let T be the linear operator on X corresponding to the matrix (a_{ij}). Let $f_1, \ldots, f_n$ denote the coefficient functionals on X; that is, $f_k(\mathbf{x}) = x_k$, where $\mathbf{x} = (x_1, \ldots x_n)$. We show that we can choose a semi-inner product that determines the norm $\| \ \|_\infty$ and for which

$$W(T) \subset \bigcup_{1 \leq k \leq n} G_k, \qquad (3.13)$$

where $G_1, \ldots, G_n$ are the Gershgorin discs for the matrix (a_{ij}).

Given $\mathbf{x} \in S(X)$, we choose an integer $k(\mathbf{x}) \in \{1, 2, \ldots, n\}$ such that $|f_{k(\mathbf{x})}(\mathbf{x})| = 1$. We then define a mapping $\psi: S(X) \to S(X')$ by taking

$$\psi(\mathbf{x}) = f_{k(\mathbf{x})}(\mathbf{x})^* f_{k(\mathbf{x})} \quad \text{for } \mathbf{x} \in S(X).$$

Then $\psi(\mathbf{x})(\mathbf{x}) = 1$, and so we can use ψ to construct a semi-inner product that determines $\| \ \|_\infty$ and for which

$$\begin{aligned} W(T) &= \{\psi(\mathbf{x})(T\mathbf{x}) : \mathbf{x} \in S(X)\} \\ &= \{f_{k(\mathbf{x})}(\mathbf{x})^* f_{k(\mathbf{x})}(T\mathbf{x}) : \mathbf{x} \in S(X)\}. \end{aligned}$$

Given $\lambda \in W(T)$, we have $\lambda = f_{k(\mathbf{x})}(x)^* f_{k(\mathbf{x})}(T\mathbf{x})$ for some $\mathbf{x} \in S(X)$. Let $k = k(\mathbf{x})$. Then

$$f_k(T\mathbf{x}) = a_{kk} f_k(\mathbf{x}) + \sum_{j \neq k} a_{kj} x_j,$$

$$\lambda = f_k(\mathbf{x})^* f_k(T\mathbf{x}) = a_{kk} + f_k(\mathbf{x})^* \sum_{j \neq k} a_{kj} x_j,$$

$$|\lambda - a_{kk}| \leq \left| \sum_{j \neq k} a_{kj} x_j \right| \leq \sum_{j \neq k} |a_{kj}| = r_k.$$

Thus $\lambda \in G_k$ and (3.13) is proved. Finally, the eigenvalues of T belong to $W(T)$, and are therefore contained in the union of the Gershgorin discs.

A similar result can be proved for operators on the space (c_0) of complex sequences that converge to zero. Let e_n denote the sequence with 1 in the nth place and 0 in all other places. Let T be a bounded linear operator on (c_0) and let (a_{ij}) denote its matrix relative to the basis $\{e_n\}$. Then we have an infinite sequence $\{G_k\}$ of Gershgorin discs

$$G_k = \{z \in \mathbb{C} : |z - a_{kk}| \leq r_k\},$$

where $r_k = \sum_{j=1}^{\infty} |a_{kj}| - |a_{kk}|$. As in the finite dimensional case, we can construct a semi-inner product determining the norm of (c_0) with

$$W(T) \subset \bigcup_{1 \leq k < \infty} G_k.$$

It then follows that the eigenvalues of T are contained in the union of the Gershgorin discs, and $\partial \operatorname{Sp}(T)$ is contained in the closure of this union.

EXAMPLE 4: Let (a_{ij}) be an $n \times n$ *stochastic matrix*; that is,

$$a_{ij} \geq 0 \quad (i,j=1,\ldots,n), \qquad \sum_{j=1}^{n} a_{ij} = 1 \quad (i=1,\ldots,n),$$

and let T be the corresponding linear operator on $\mathbb{C}^n$. We prove that all eigenvalues λ of T with $|\lambda|=1$ have ascent 1. To do this, we give $\mathbb{C}^n$ the norm $\| \ \|_\infty$ as in the last example. Given $\mathbf{x} = (x_1, \ldots x_n)$, we have

$$|(T\mathbf{x})_i| = \left| \sum_{j=1}^{n} a_{ij} x_j \right| \leq \sum_{j=1}^{n} a_{ij} \|\mathbf{x}\| = \|\mathbf{x}\|,$$

and therefore

$$\|T\mathbf{x}\| \leq \|\mathbf{x}\|.$$

It follows that for any printer Φ, $\operatorname{cl co} \Phi(T)$ is contained in the closed unit disc, and therefore every eigenvalue λ with $|\lambda|=1$ is a boundary point of $\operatorname{cl co} \Phi(T)$. Therefore λ has ascent 1 by Theorem 6.

EXAMPLE 5: Let T be an operator on a finite dimensional space X such that for some choice of a printer Φ on X, the set $\Phi(T)$ is contained in a straight line segment. Then X has a basis relative to which T is represented as a diagonal matrix. For $\operatorname{cl co} \Phi(T)$ is contained in the same straight line segment, and therefore all eigenvalues of T are boundary points of $\operatorname{cl co} \Phi(T)$. Therefore by Theorem 6, all eigenvalues of T have ascent one. Therefore the classical canonical form of T is a diagonal matrix.

The same conclusion follows, by Crabb's extension of Theorem 6, whenever $V(T)$ has void interior for some choice of the norm.

4. THE SPATIAL NUMERICAL RANGE

In this section, X will denote a complex Banach space. We have seen in §3 that if $W(T)$ is the numerical range corresponding to a semi-inner product that determines the norm of X, then we have

$$\operatorname{pSp}(T) \subset W(T), \qquad \partial \operatorname{Sp}(T) \subset \operatorname{cl} W(T).$$

The following example shows that we cannot hope to prove that $\operatorname{Sp}(T) \subset \operatorname{cl} W(T)$, in general.

EXAMPLE 6: Let $\mathbb{D} = \{z \in \mathbb{C} : |z| < 1\}$, $\mathbb{E} = \{z \in \mathbb{C} : |z| \leq 1\}$, $\mathbb{T} = \{z \in \mathbb{C} : |z| = 1\}$, and let X denote the *disc algebra* of all continuous complex functions on $\mathbb{E}$ that are analytic in $\mathbb{D}$. Then X is a Banach space with respect to the uniform norm.

Given $x \in S(X)$, the maximum modulus theorem implies that there exists a point $z_x \in \mathbb{T}$ such that $|x(z_x)| = 1$. We choose such a point $z_y \in \mathbb{T}$ corresponding to each $y \in S(X)$ and define $\psi(y) \in S(X')$ by taking

$$\psi(y)(x) = y(z_y)^* x(z_y) \quad \text{for } x \in X.$$

Then $\psi(y)(y) = 1$, and corresponding to ψ, we have a semi-inner product and hence a numerical range $W(T)$ given by

$$W(T) = \{ y(z_y)^* (Ty)(z_y) : y \in S(X) \}.$$

In particular, let T be the multiplication operator on X given by

$$(Tx)(z) = zx(z) \quad \text{for } z \in \mathbb{E}, x \in X.$$

Then we have

$$W(T) = \{ y(z_y)^* z_y y(z_y) : y \in S(X) \}$$
$$= \{ z_y : y \in S(X) \} = \mathbb{T},$$

the last equality because, for each $z \in \mathbb{T}$, there exists $y \in S(X)$ with modulus 1 only at z.

On the other hand it is easy to check that $\operatorname{Sp}(T) = \mathbb{E}$. For, if $\lambda \in \mathbb{E} \setminus \operatorname{Sp}(T)$, there exists a bounded linear operator S on X with $(\lambda - T)S = I$. Take $\phi = S1$, where 1 is the function with the constant value 1; then $(\lambda - T)S1 = I1$ yields $(\lambda - z)\phi(z) = 1$ for $z \in \mathbb{E}$, which is absurd.

In this example $\operatorname{cl} W(T) = W(T) = \partial \operatorname{Sp}(T)$, and $\operatorname{Sp}(T)$ has the non-trivial interior $\mathbb{D}$ which is absent from $\operatorname{cl} W(T)$.

For the spatial numerical range $V(T)$, more has been proved. An important theorem of Bishop and Phelps (1961) on the geometry of Banach spaces tells us that the set of functionals $f \in S(X')$ such that $f \in D(x)$ for some $x \in S(X)$ is norm dense in $S(X')$. Using this theorem, Williams (1967) proved that

$$\operatorname{Sp}(T) \subset \operatorname{cl} V(T).$$

The spatial numerical range and its closure are not in general convex, so it was not to be expected that $\operatorname{cl} V(T)$ would contain the convex hull of the spectrum. Nevertheless, this was proved by Chr. Zenger (1968) for finite dimensional spaces X and extended to infinite dimensional spaces by Crabb (1969). Zenger's proof involves such an ingenious, though elementary, idea that we give it in full.

LEMMA 2: *Let $\alpha_1, \alpha_2, \ldots, \alpha_n$ be positive real numbers with $\alpha_1 + \alpha_2 + \cdots + \alpha_n = 1$, let $\| \ \|$ be a norm on the linear space $\mathbb{C}^n$, and let U denote the closed unit ball in $\mathbb{C}^n$ with respect to this norm. Then there exists $(w_1, w_2, \ldots, w_n) \in U$ with $w_1 w_2 \cdots w_n \neq 0$ such that*

$$|\alpha_1 w_1^{-1} z_1 + \alpha_2 w_2^{-1} z_2 + \cdots + \alpha_n w_n^{-1} z_n| \leq 1$$

for all $(z_1, \ldots, z_n) \in U$.

Proof: Define a function F on $\mathbb{C}^n$ by $F(\mathbf{z}) = |z_1|^{\alpha_1} |z_2|^{\alpha_2} \cdots |z_n|^{\alpha_n}$, where $\mathbf{z} = (z_1, z_2, \ldots, z_n)$. Since U is the closed unit ball in a finite dimensional space it is compact, and the continuous function F attains its supremum on U; thus there exists $\mathbf{w} \in U$ such that

$$F(\mathbf{z}) \leq F(\mathbf{w}) \quad \text{for } \mathbf{z} \in U. \tag{4.1}$$

For sufficiently small $\epsilon > 0$, the point $(\epsilon, \epsilon, \ldots, \epsilon)$ belongs to U, and so

$$0 < \epsilon = \epsilon^{\alpha_1} \epsilon^{\alpha_2} \cdots \epsilon^{\alpha_n} \leq F(\mathbf{w}) = |w_1|^{\alpha_1} \cdots |w_n|^{\alpha_n}.$$

Since $\alpha_k > 0$, this proves that $w_k \neq 0$ for $1 \leq k \leq n$, and therefore (4.1) gives

$$\left|\frac{z_1}{w_1}\right|^{\alpha_1} \left|\frac{z_2}{w_2}\right|^{\alpha_2} \cdots \left|\frac{z_n}{w_n}\right|^{\alpha_n} \leq 1 \quad \text{for all } \mathbf{z} \in U. \tag{4.2}$$

Given $\mathbf{z} \in U$ and $0 \leq \lambda \leq 1$, we have $\lambda \mathbf{z} + (1-\lambda)\mathbf{w} \in U$, and so (4.2) implies that

$$\left|\lambda \frac{z_1}{w_1} + 1 - \lambda\right|^{\alpha_1} \left|\lambda \frac{z_2}{w_2} + 1 - \lambda\right|^{\alpha_2} \cdots \left|\lambda \frac{z_n}{w_n} + 1 - \lambda\right|^{\alpha_n} \leq 1.$$

Therefore

$$\alpha_1 \log\left|1 + \lambda\left(\frac{z_1}{w_1} - 1\right)\right| + \cdots + \alpha_n \log\left|1 + \lambda\left(\frac{z_n}{w_n} - 1\right)\right| \leq 0.$$

We have

$$1 + \lambda\left(\operatorname{Re}\left(\frac{z_k}{w_k} - 1\right)\right) = \operatorname{Re}\left(1 + \lambda\left(\frac{z_k}{w_k} - 1\right)\right) \leq \left|1 + \lambda\left(\frac{z_k}{w_k} - 1\right)\right|$$

and as $\lambda \to 0$,

$$\log\left(1 + \lambda\left(\operatorname{Re}\left(\frac{z_k}{w_k}\right) - 1\right)\right) = \lambda\left(\operatorname{Re}\left(\frac{z_k}{w_k}\right) - 1\right) + O(\lambda^2).$$

Therefore, dividing by λ and then letting $\lambda \to 0$, we obtain

$$\alpha_1\left(\operatorname{Re}\left(\frac{z_1}{w_1}\right) - 1\right) + \cdots + \alpha_n\left(\operatorname{Re}\left(\frac{z_n}{w_n}\right) - 1\right) \leq 0,$$

$$\operatorname{Re}(\alpha_1 w_1^{-1} z_1 + \cdots + \alpha_n w_n^{-1} z_n) \leq \alpha_1 + \alpha_2 + \cdots + \alpha_n = 1.$$

We have proved that the linear functional on $\mathbb{C}^n$ defined by

$\phi(\mathbf{z}) = \alpha_1 w_1^{-1} z_1 + \cdots + \alpha_n w_n^{-1} z_n$ satisfies

$$\operatorname{Re}\phi(\mathbf{z}) \leq 1 \quad \text{for all } \mathbf{z} \in U. \tag{4.3}$$

Given $\mathbf{z} \in U$, choose a real number θ such that $e^{i\theta}\phi(\mathbf{z}) = |\phi(\mathbf{z})|$. Then $e^{i\theta}\mathbf{z} \in U$, and so (4.3) gives

$$|\phi(\mathbf{z})| = \operatorname{Re} e^{i\theta}\phi(\mathbf{z}) = \operatorname{Re}\phi(e^{i\theta}\mathbf{z}) \leq 1.$$

THEOREM 8: *Let $a_1, \ldots, a_r$ be linearly independent elements of X, and let $\alpha_k \geq 0$ ($1 \leq k \leq r$) and $\sum_{k=1}^{r} \alpha_k = 1$. Then there exist complex numbers $w_1, \ldots, w_r$ such that $w_1 a_1 + \cdots + w_r a_r = x \in S(X)$ and there exists $f \in D(x)$ such that $f(w_k a_k) = \alpha_k$ for $k = 1, \ldots, r$.*

Proof: We may suppose without loss of generality that there is a positive integer n with $1 \leq n \leq r$ such that $\alpha_k > 0$ ($1 \leq k \leq n$) and $\alpha_k = 0$ ($n < k \leq r$). Define a mapping $P: \mathbb{C}^n \to X$ by $P\mathbf{z} = z_1 a_1 + \cdots + z_n a_n$ for all $\mathbf{z} = (z_1, \ldots, z_n) \in \mathbb{C}^n$. Then $P\mathbf{z} \neq 0$ whenever $\mathbf{z} \neq 0$, and so we can define a norm $\|\ \|$ on $\mathbb{C}^n$ by taking $\|\mathbf{z}\| = \|P\mathbf{z}\|$. Take $Y = P\mathbb{C}^n$; then P is a linear isometry of $\mathbb{C}^n$ onto Y and so it has an inverse $P^{-1}: Y \to \mathbb{C}^n$ which is also a linear isometry. Let $\mathbf{w} \in \mathbb{C}^n$ be as in Lemma 2; define a linear functional ϕ on $\mathbb{C}^n$ by

$$\phi(\mathbf{z}) = \alpha_1 \frac{z_1}{w_1} + \cdots + \alpha_n \frac{z_n}{w_n},$$

and define f_0 on Y by $f_0(y) = \phi(P^{-1}y)$ for $y \in Y$. We have $|f_0(y)| \leq \|P^{-1}y\| = \|y\|$ for $y \in Y$, so that f_0 is a linear functional on Y with $\|f_0\| \leq 1$. Also $\|P\mathbf{w}\| = \|\mathbf{w}\| \leq 1$ and

$$f_0(P\mathbf{w}) = \phi(\mathbf{w}) = \alpha_1 + \cdots + \alpha_n = 1.$$

Therefore $\|f_0\| = \|P\mathbf{w}\| = 1$.

By the Hahn-Banach theorem, there exists a linear functional f on X with $f|Y = f_0$ and $\|f\| = \|f_0\| = 1$. Take $x = P\mathbf{w}$, that is, $x = w_1 a_1 + \cdots + w_n a_n$. Then $\|x\| = 1$ and $f(x) = f_0(P\mathbf{w}) = 1$. Thus $f \in D(x)$.

Since, for all $z \in \mathbb{C}^n$,

$$f_0(z_1a_1 + \cdots + z_na_n) = f_0(Pz) = \phi(z) = \frac{\alpha_1}{w_1}z_1 + \cdots + \frac{\alpha_n}{w_n}z_n,$$

we have $f_0(a_k) = \alpha_k/w_k$. Therefore $f(w_ka_k) = \alpha_k$ for $k = 1,\ldots,n$. Finally, take $w_k = 0$ for $k = n+1,\ldots,r$.

COROLLARY: *The convex hull of the eigenvalues is contained in the spatial numerical range.*

Proof: Let $\lambda_1,\ldots,\lambda_r$ be distinct eigenvalues of T and let $\alpha_k \geq 0$ $(1 \leq k \leq r)$ satisfy $\sum_{k=1}^{r} \alpha_k = 1$. We have to prove that

$$\alpha_1\lambda_1 + \cdots + \alpha_r\lambda_r \in V(T). \tag{4.4}$$

There exist eigenvectors $a_1,\ldots,a_r$ with $a_k \neq 0$, $Ta_k = \lambda_k a_k$ $(1 \leq k \leq r)$. The vectors $a_1,\ldots,a_r$ are linearly independent, and so Theorem 8 gives complex numbers $w_1,\ldots,w_r$ such that $x = w_1a_1 + \cdots + w_ra_r \in S(X)$ and $f \in D(x)$ such that $f(w_ka_k) = \alpha_k$ for $k = 1,\ldots,r$. Then

$$f(Tx) = \sum_{k=1}^{r} f(Tw_ka_k) = \sum_{k=1}^{r} \lambda_k f(w_ka_k) = \sum_{k=1}^{r} \alpha_k\lambda_k.$$

Since $f \in D(x)$, then $f(Tx) \in V(T)$ and (4.4) is proved.

Crabb (1969) proved that the convex hull of the spectrum of T is contained in the closure of the spatial numerical range of T; that is,

$$\operatorname{co}\operatorname{Sp}(T) \subset \operatorname{cl} V(T).$$

The proof has three ingredients; the fact that $\partial \operatorname{Sp}(T) \subset \operatorname{ap}\operatorname{Sp}(T)$, a classical theorem of Carathéodory on convex sets in finite dimensional spaces, according to which each point of $\operatorname{co}\operatorname{Sp}(T)$ is a convex combination of at most three points of $\partial \operatorname{Sp}(T)$, and Zenger's theorem. The ingredients are put together with an approximation argument. (For the details, see NR II, pp. 19–20.)

It is known that $V(T)$ is a connected set, and it is obviously bounded. When the space has finite dimension, then $V(T)$ is compact, but this need not be true when the dimension is infinite even for Hilbert spaces. It would be very interesting to know more about the topology and geometry of $V(T)$ and $\operatorname{cl} V(T)$. In particular, we do not know any example in which the complement of $\operatorname{cl} V(T)$ has more than one component. In every example that we know, each point of $\partial V(T)$ is the endpoint of an open half-line contained in the complement of $\operatorname{cl} V(T)$.

The numerical range $W(T)$ corresponding to a semi-inner product need not be connected. Let K be a non-void compact subset of $\mathbb{C}$, let X denote the space $C(K)$ of continuous complex functions on K, and let T be the multiplication operator on X given by $(Tx)(z) = zx(z)$ for $z \in K$, $x \in X$. A construction of a semi-inner product as in Example 2 gives $W(T) = K$. Thus every non-void compact set in $\mathbb{C}$ is the numerical range $W(T)$ of some operator T. We do not know whether every non-void bounded set in $\mathbb{C}$ can occur as a numerical range $W(T)$.

We end this section with an interesting theorem due to Williams (1967) on the spectrum of a product of operators.

THEOREM 9: *Let* $S, T \in BL(X)$ *and let* $0 \notin \operatorname{cl} V(T)$. *Then* $\operatorname{Sp}(T^{-1}S) \subset \{\lambda^{-1}\mu : \lambda \in \operatorname{cl} V(T), \mu \in \operatorname{cl} V(S)\}$.

Proof: Since $0 \notin \operatorname{cl} V(T)$, we have $0 \notin \operatorname{Sp}(T)$, and so T is invertible. Let ζ be a complex number not of the form $\lambda^{-1}\mu$ with $\lambda \in \operatorname{cl} V(T)$, $\mu \in \operatorname{cl} V(S)$. Then there exists $\delta > 0$ such that $|\zeta\lambda - \mu| \geq \delta$ for all $\lambda \in V(T)$ and $\mu \in V(S)$. Given $x \in S(X)$ and $f \in D(x)$, we have

$$|f((\zeta T - S)x)| = |\zeta f(Tx) - f(Sx)| \geq \delta,$$

since $f(Tx) \in V(T)$ and $f(Sx) \in V(S)$. Therefore $\|(\zeta T - S)x\| \geq \delta$, and $\|(\zeta T - S)^*f\| \geq \delta$; the first of these inequalities holds for all $x \in S(X)$ and the second also holds for all $f \in S(X')$ since the Bishop-Phelps theorem allows us to approximate arbitrary $f \in S(X')$ by $g \in D(x)$ for some $x \in S(X)$. We now have $\|(\zeta T - S)x\|$

$\geqslant \delta \|x\|$ for all $x \in X$ and $\|(\zeta T - S)^* f\| \geqslant \delta \|f\|$ for all $f \in X'$; it follows that $\zeta T - S$ is invertible in $BL(X)$. Finally

$$\zeta I - T^{-1}S = T^{-1}(\zeta T - S),$$

and so $\zeta I - T^{-1}S$ is invertible, or $\zeta \notin \mathrm{Sp}(T^{-1}S)$.

5. HERMITIAN OPERATORS ON BANACH SPACES

In this section X denotes a complex Banach space. In §3, we defined an operator T on X to be Hermitian if $\Phi(T) \subset \mathbb{R}$ for some (and therefore all) printers Φ on X, and we proved that the Hermitian operators are precisely those for which

$$\|\exp(itT)\| = 1 \qquad (t \in \mathbb{R}). \tag{5.1}$$

For operators on Hilbert space, we have $\|T^*T\| = \|T\|^2$. If T is self-adjoint, this gives $\|T^2\| = \|T\|^2$, and by iteration, $\|T^{2^n}\| = \|T\|^{2^n}$. Therefore self-adjoint operators on Hilbert space satisfy $r(T) = \|T\|$, where (as in §1) $r(T)$ denotes the spectral radius of T. It is a remarkable fact, due to Sinclair (1971), that this equality holds also for Hermitian operators on Banach spaces.

THEOREM 10: *Let T be a Hermitian operator on X. Then $r(T) = \|T\|$.*

Proof: For all operators T we have $r(T) \leqslant \|T\|$. By homogeneity, we may assume that $r(T) < \pi/2$, and then it is enough to prove that $\|T\| \leqslant \pi/2$. We have $\sin T = (1/2i)(\exp(iT) - \exp(-iT))$ and so (5.1) gives

$$\|\sin T\| \leqslant 1. \tag{5.2}$$

The idea of the proof is to consider the usual power series for $\arcsin z$:

$$\arcsin z = z + \tfrac{1}{2} \cdot \tfrac{1}{3} z^3 + \cdots = \sum_{r=1}^{\infty} c_r z^r. \tag{5.3}$$

Here we have $c_r \geq 0$ for $r = 1, 2, \ldots$, and

$$t = \sum_{r=1}^{\infty} c_r (\sin t)^r \quad \text{for all } t \in [-\pi/2, \pi/2], \tag{5.4}$$

from which, in particular, we have $\sum_{r=1}^{\infty} c_r = \pi/2$. If we can prove that

$$T = \sum_{r=1}^{\infty} c_r (\sin T)^r, \tag{5.5}$$

we shall have, by (5.2), the inequality

$$\|T\| \leq \sum_{r=1}^{\infty} c_r \|\sin T\|^r \leq \sum_{r=1}^{\infty} c_r = \pi/2,$$

and the theorem will be proved.

To prove (5.5) we use the functional calculus for functions analytic on a neighborhood of the spectrum. Since T is Hermitian, we have $\mathrm{Sp}(T) \subset \mathrm{cl}\, V(T) \subset \mathbb{R}$, and we are assuming that $\sup|\mathrm{Sp}(T)| < \pi/2$. Thus $\mathrm{Sp}(T)$ is a compact subset of the open real interval $(-\pi/2, \pi/2)$. By the Weierstrass M-test, the series (5.3) converges uniformly when $|z| \leq 1$, and so it defines a function f analytic on the open unit disc $\mathbb{D}$. By continuity of the mapping $z \mapsto \sin z$, there exists an open connected neighborhood G of $\mathrm{Sp}(T)$ such that $\sin G \subset \mathbb{D}$. By (5.4) and analyticity, we have $f(\sin z) = z$ for all $z \in G$. We now apply the functional calculus (see for example Bonsall and Duncan [1, Theorem 7.6]) to obtain

$$f(\sin T) = T.$$

Thus (5.5) is proved, and the proof is complete.

Sinclair actually proved a stronger theorem, namely, that for Hermitian operators T and all complex numbers α, β, we have

$$r(\alpha I + \beta T) = \|\alpha I + \beta T\|. \tag{5.6}$$

For a compact self-adjoint operator T on a Hilbert space there is the classical spectral decomposition

$$T = \Sigma \lambda_n P_n,$$

where $\{\lambda_n\}$ is the sequence of non-zero eigenvalues of T, P_n is the spectral projection corresponding to λ_n, and the series converges in norm. A similar result holds for Hermitian operators on Banach spaces provided that the sequence $\{\lambda_n\}$ converges to zero sufficiently rapidly. The following theorem is far from being the strongest possible, but it is easy to prove.

THEOREM 11: *Let T be a compact Hermitian operator on X. Let $\{\lambda_n\}$ denote the distinct non-zero eigenvalues of T arranged so that*

$$|\lambda_{n+1}| \leq |\lambda_n| \quad \text{for } n=1,2,\ldots,$$

and suppose (if there are infinitely many of them) that $\lim_{n\to\infty} n\lambda_n = 0$. Let P_n be the spectral projection corresponding to the eigenvalue λ_n. Then
  (i) *each eigenvalue has ascent 1;*
  (ii) $\|P_n\| = 1$ *for $n = 1, 2, \ldots$;*
  (iii) $T = \Sigma \lambda_n P_n$, *with the series convergent in the operator norm.*

Proof: (i) Since $V(T) \subset \mathbb{R}$, each eigenvalue of T is a boundary point of $\operatorname{cl}\operatorname{co} V(T)$. Therefore, by Theorem 6, it has ascent 1.

(ii) Since λ_n has ascent 1, $P_n X$ is the kernel and $(I - P_n)X$ is the range of $\lambda_n I - T$. Since 0 is a boundary point of $\operatorname{cl}\operatorname{co} V(\lambda_n I - T)$, the corollary to Theorem 7 shows that $P_n X \perp (I - P_n)X$; that is,

$$\|P_n x + (I - P_n)y\| \geq \|P_n x\| \quad \text{for all } x, y \in X.$$

If we take $y = x$, this gives $\|P_n x\| \leq \|x\|$ and proves (ii).

(iii) Suppose that there are infinitely many distinct eigenvalues λ_n, and let $S_n = \Sigma_{k=1}^n \lambda_k P_k$ and $R_n = \Sigma_{k=1}^n P_k$. We have

$$(\lambda_j - \lambda_i)P_j P_i X = P_j(\lambda_j I - T)P_i X \subset P_j(I - P_j)X = \{0\},$$

and so $P_j P_i = 0$ for $j \neq i$. Thus R_n is a projection and

$$(I - R_n)X = \bigcap_{k=1}^n (I - P_n)X,$$

which shows that $(I - R_n)X$ is an invariant subspace for T. Let T_n denote the restriction of T to this subspace. Then $V(T_n) \subset V(T) \subset$

ℝ, which shows that T_n is a Hermitian operator. Therefore, by Sinclair's theorem (Theorem 10), we have $\|T_n\| = r(T_n)$. But $\text{Sp}(T_n) = \{0, \lambda_{n+1}, \lambda_{n+2}, \ldots\}$, and so $\|T_n\| = |\lambda_{n+1}|$. We have the relation $T - S_n = T_n(I - R_n)$, and by (ii), $\|R_n\| \leq n$. Therefore,

$$\|T - S_n\| \leq (n+1)\|T_n\| = (n+1)|\lambda_{n+1}|.$$

This gives $\lim_{n \to \infty} \|T - S_n\| = 0$ and completes the proof when there are infinitely many distinct eigenvalues. If there are only n non-zero eigenvalues, the above argument gives $\|T_n\| = r(T_n) = 0$ and $T = S_n$.

Using an ingenious but still elementary argument, B. Bollobás [5] has proved that the conclusion of this theorem holds under the weaker hypothesis:

$$\lim_{n \to \infty} n^{\frac{1}{2}} \lambda_n = 0,$$

and that $\frac{1}{2}$ is the best possible power of n; in fact, that $\sup_n n^{\frac{1}{2}} |\lambda_n| < \infty$ is not enough.

If we replace convergence of the series $\Sigma \lambda_n P_n$ by Fejér-Bochner summation of the same series, the theorem holds without restriction on the eigenvalues. This was proved for weakly complete Banach spaces by Yu. I. Lyubič [Ju. I. Ljubič] (1960) (see NR II, §28) and extended to general Banach spaces by Bollobás [5]. A key step in the proof is the fact that for each $x \in X$ and $f \in X'$, the mapping $t \mapsto f((\exp itT)x)$ of ℝ into ℂ is uniformly almost periodic on ℝ, whenever T is a compact Hermitian operator on X.

6. THE ALGEBRA NUMERICAL RANGE

In §3 we defined the algebra numerical range of a bounded linear operator. We consider next numerical ranges in the setting of normed algebras. For convenience of exposition we shall suppose hereafter that A is a complex Banach algebra which is also unital (that is, A has a unit element 1 and $\|1\| = 1$). As in §2, we write

$$D(1) = \{f \in S(A') : f(1) = 1\}.$$

The elements of $D(1)$ are called *normalized states*.

DEFINITION: The *algebra numerical range* $V(a)$ is defined, for each $a \in A$, by $V(a) = \{f(a) : f \in D(1)\}$.

Since $D(1)$ is a non-empty weak* compact convex subset of A', it is clear that $V(a)$ is a non-empty compact convex subset of $\mathbb{C}$. Evidently $V(\alpha + \beta a) = \alpha + \beta V(a)$ for $\alpha, \beta \in \mathbb{C}$, and $\sup |V(a)| \leq \|a\|$.

Since $V(a)$ is a compact convex subset of $\mathbb{C}$, it is the intersection of all the closed discs that contain it. In fact an application of the Hahn-Banach theorem gives the more precise (and useful) representation

$$V(a) = \bigcap_{z \in \mathbb{C}} \{\zeta : |z - \zeta| \leq \|z - a\|\}. \tag{6.1}$$

This also shows that $V(a)$ is completely determined by the values of the norm on the linear span of 1 and a.

It should be noted that $V(a)$ is the spatial numerical range $V(T_a)$ where $T_a : A \to A$ is defined by $T_a x = ax$ ($x \in A$). We have $V(a) \subset V(T_a)$ since $1 \in S(A)$. On the other hand, given $x \in S(A)$, $f \in D(x)$, define $g(y) = f(yx)$ ($y \in A$). Then $g \in D(1)$ and $g(a) = f(ax) = f(T_a x)$. As simple corollaries of Lemma 1 and Theorem 5 we thus have the following formulae for the supremum of the real parts of the elements of $V(a)$:

$$\max \operatorname{Re} V(a) = \inf_{t > 0} t^{-1}(\|1 + ta\| - 1)$$

$$= \lim_{t \to 0+} t^{-1}(\|1 + ta\| - 1), \tag{6.2}$$

$$\max \operatorname{Re} V(a) = \sup_{t > 0} t^{-1} \log \|\exp(ta)\|$$

$$= \lim_{t \to 0+} t^{-1} \log \|\exp(ta)\|. \tag{6.3}$$

THEOREM 12: *If $a \in A$, then $\|a\| \leq e \max |V(a)|$.*

Proof: Let $\max |V(a)| \leq 1$. By (6.3) we have $\|\exp(za)\| \leq e^{|z|}$ for $z \in \mathbb{C}$. Let Γ denote the unit circle with positive orientation. Since

$$a = \frac{1}{2\pi i} \int_\Gamma \exp(za) \frac{dz}{z^2},$$

it follows that $\|a\| \leq e$, as required.

In particular we have $a=0$ whenever $V(a)=\{0\}$; this shows that $D(1)$ contains enough functionals to separate the points of A. A point $x \in S(A)$ is called a *smooth point* if $D(x)$ is a singleton, and is called a *vertex* of the closed unit ball if $D(x)$ separates the points of A. Thus the unit element is a vertex of the closed unit ball, and it is a smooth point only when $A = \mathbb{C}1$.

The fact that $\mathrm{Sp}(a) \subset V(a)$ is much more elementary than the corresponding result of J. P. Williams for the spatial numerical range.

LEMMA 3: *If $a \in A$, then $\mathrm{Sp}(a) \subset V(a)$.*

Proof: Let $\lambda \in \mathbb{C} \setminus V(a)$. By (6.1) there exists $z \in \mathbb{C}$ such that $|z-\lambda| > \|z-a\|$. Therefore $\|(z-\lambda)^{-1}(z-a)\| < 1$, so the element $1-(z-\lambda)^{-1}(z-a)$ has an inverse in A. It follows that $\lambda - a$ has an inverse in A and so $\lambda \in \mathbb{C} \setminus \mathrm{Sp}(a)$. The proof is complete.

Since $V(a)$ is convex, the above lemma implies that $\mathrm{co\,Sp}(a) \subset V(a)$. We saw in §1 that if T is a bounded linear operator on a Hilbert space, then corners of $\mathrm{cl}\,W(T)$ belong to $\mathrm{Sp}(T)$. To obtain a generalization to the setting of Banach algebras we need to restrict the angle at the corner. The theorem is due independently to B. Schmidt [10] and A. M. Sinclair.

THEOREM 13: *Let λ be a corner of $V(a)$ of angle less than $\pi/2$ radians. Then $\lambda \in \mathrm{Sp}(a)$.*

Proof: By rotation and translation we may assume, without loss of generality, that $\lambda = 0$ and that for some $\gamma > \pi/4$,

$$\max \mathrm{Re}\, V(e^{i\theta}a) < 0 \quad \text{for } |\theta| \leq \gamma. \tag{6.4}$$

By (6.3) we have $\|\exp(te^{i\theta}a)\| \leq 1$ for $t \geq 0$ and $|\theta| \leq \gamma$. Since $0 \in V(a)$, there exists $f \in D(1)$ with $f(a) = 0$. Thus

$$1 \geq |f(\exp(te^{i\theta}a))| = |1 + \tfrac{1}{2}t^2 e^{2i\theta} f(a^2) + \cdots|$$
$$\geq |1 + \tfrac{1}{2}t^2 e^{2i\theta} f(a^2)| - O(t^3) \quad (t \to 0+).$$

This gives $\operatorname{Re}(e^{2i\theta}f(a^2)) \leq 0$ for all $|\theta| \leq \gamma$, and so $f(a^2) = 0$. The above reasoning now gives

$$1 \geq |1 + \tfrac{1}{6}t^3 e^{3i\theta}f(a^3)| - O(t^4) \qquad (t \to 0+)$$

and hence $f(a^3) = 0$. By induction we obtain $f(a^n) = 0$ for all $n = 1, 2, 3, \ldots$ and so

$$\|\exp(ta)\| \geq |f(\exp(ta))| = 1 \qquad (t > 0).$$

From elementary Banach algebra theory it is known that

$$\begin{aligned}\max \operatorname{Re} \operatorname{Sp}(a) &= \inf_{t>0} t^{-1} \log \|\exp(ta)\| \\ &= \lim_{t \to \infty} t^{-1} \log \|\exp(ta)\|.\end{aligned} \qquad (6.5)$$

Therefore $\max \operatorname{Re} \operatorname{Sp}(a) \geq 0$. But $\operatorname{Sp}(a) \subset V(a)$, and so it follows from (6.4) that $0 \in \operatorname{Sp}(a)$.

The above theorem gives an entirely elementary proof of the fact that $V(h) = \operatorname{co} \operatorname{Sp}(h)$ for any Hermitian element h. The theorem is best possible, as the following example shows.

EXAMPLE 8: Let $A = BL(X)$, where X is $\mathbb{C}^2$ with the norm given by $\|(z, w)\| = \max\{|z|, |w|, (1/\sqrt{2})|z - iw|\}$ and let

$$a = \begin{pmatrix} 1 & 0 \\ 0 & 0 \end{pmatrix}.$$

It is evident that $\operatorname{Sp}(a) = \{0, 1\}$, and with sufficient industry, it may be verified that $V(a) = \operatorname{co}\{0, 1, \tfrac{1}{2}(1+i), \tfrac{1}{2}(1-i)\}$.

7. THE COMPLEX LINEAR SPAN OF THE HERMITIAN ELEMENTS

We recall that an element a of A is Hermitian if $V(a) \subset \mathbb{R}$. The set of all Hermitian elements of A is denoted by $\operatorname{Her}(A)$. Evidently $\operatorname{Her}(A)$ is a closed real linear subspace of A. In general, Hermitian elements are conspicuous by their absence. For example, if A is the convolution algebra $l^1(G)$, where G is any non-trivial discrete group, then $\text{Her}(A) = \mathbb{R}1$. At the other extreme we have already noted that if $A = BL(H)$ then $a \in \text{Her}(A)$ if and only if $a^* = a$; and in this case we have that

$$a \in \text{Her}(A) \Rightarrow a^2 \in \text{Her}(A). \tag{7.1}$$

Property (7.1) fails in general, as we shall see in §8. It is a remarkable fact that property (7.1) is a necessary and sufficient condition that the elements of $\text{Her}(A)$ may be identified as self-adjoint operators on some Hilbert space. We shall indicate some of the steps involved in this identification; full details may be found in NR I and II.

Let $J(A) = \text{Her}(A) + i\text{Her}(A)$. Note that each element of $J(A)$ has a unique representation of the form $h + ik$ with $h, k \in \text{Her}(A)$; for if $h + ik = 0$ with $h, k \in \text{Her}(A)$, then $V(h) = -iV(k) \subset \mathbb{R} \cap i\mathbb{R} = \{0\}$ and so $h = k = 0$, by Theorem 12. We may thus define a mapping * from $J(A)$ to itself by

$$(h + ik)^* = h - ik \quad \text{for } h, k \in \text{Her}(A).$$

It is readily verified that $J(A)$ is a complex Banach space and * is a continuous linear involution on $J(A)$.

The space $\text{Her}(A)$ is closed under the imaginary Jordan product:

$$(h, k) \mapsto i(hk - kh),$$

but it need not be closed under the real Jordan product:

$$(h, k) \mapsto hk + kh.$$

The first statement follows from (6.2) and consideration of the identity

$$\|\exp(ith)\exp(itk)\exp(-ith)\exp(-itk)\| = 1$$

for small real t; the second is a consequence of the following lemma.

LEMMA 4: *The following statements are equivalent*:

(i) $J(A)$ *is an algebra*;
(ii) $h \in \text{Her}(A) \Rightarrow h^2 \in J(A)$;
(iii) $h \in \text{Her}(A) \Rightarrow h^2 \in \text{Her}(A)$;
(iv) $h, k \in \text{Her}(A) \Rightarrow hk + kh \in \text{Her}(A)$.

Under any of the above conditions, $J(A)$ is a Banach star algebra with continuous involution * in which $\text{Her}(A) = \{a \in A : a^* = a\}$. In particular this is the case if $J(A) = A$. Such algebras were considered by I. Vidav in his seminal paper of 1956. Vidav imposed an additional condition on A since he was unaware of the implication (i)$\Rightarrow$(iii) in the above lemma, which was established much later by T. W. Palmer (1968). Vidav proved that there exists a Hilbert space H and a bicontinuous star monomorphism $\pi : A \to BL(H)$. Subsequently, E. Berkson (1966) and B. W. Glickfeld (1966) proved independently that π is in fact an isometry. Finally Palmer established Lemma 4 and provided a simple proof that π is an isometry.

THEOREM 14: *Let $J(A) = A$. Then there exists a Hilbert space H and an isometric star monomorphism π of A into $BL(H)$*.

Theorem 14 gives an abstract characterization of C^*-algebras, that is, self-adjoint closed subalgebras of $BL(H)$. In the literature this theorem is often called the "Vidav-Palmer Theorem." We refer the reader to [1] for a detailed proof. The first main step is to establish that $V(a^*a) \subset \mathbb{R}^+$ for each $a \in A$. Thus, each normalized state is a positive functional and Theorem 12 then shows that the Gel'fand-Naĭmark-Segal star representation π is faithful. Palmer's proof that π is an isometry depends on an extension of the Russo-Dye theorem for C^*-algebras; Harris (1972) provided a very elegant proof by using the generalized Möbius transformations.

The Vidav-Palmer theorem is undoubtedly a high point in numerical range theory. It provides an entirely geometrical characterization of abstract C^*-algebras. Recall that a B^*-algebra is a Banach star algebra A such that $\|x^*x\| = \|x\|^2$ for all $x \in A$. The

Vidav-Palmer theorem asserts that, if A is a (unital) Banach algebra, the existence of an involution on A for which A is a B^*-algebra depends only on the shape of the unit ball near 1. The elementary techniques involved in the proof of the Vidav-Palmer theorem provide a simple proof of the Gel'fand-Naĭmark result that each B^*-algebra is isometrically star isomorphic to a C^*-algebra.

The Vidav-Palmer theorem also provides elegant proofs of some standard theorems. For example, a theorem of Kaplansky (1949) and Segal (1947) states that if A is a B^*-algebra and J is a closed bi-ideal of A, then $J^* = J$ and the quotient algebra A/J is a B^*-algebra. This may now be left as an easy exercise for the reader. A theorem of Glimm and Kadison (1960) states that if A is a (unital) Banach star algebra such that $\|x^*x\| = \|x^*\|\|x\|$ for all $x \in A$, then A is a B^*-algebra. By the Vidav-Palmer theorem it is enough to prove that if $h^* = h$, then $h \in \mathrm{Her}(A)$. To see this, let $\mu = \max \mathrm{Re}\, V(ih)$, $\lambda = \min \mathrm{Re}\, V(ih)$, $t > 0$. Using (6.2) we obtain

$$1 + t(\mu - \lambda) + o(t) = \|(1 + ith)^*\| \|1 + ith\|$$
$$= \|1 + t^2 h^2\|$$
$$= 1 + O(t^2) \quad (t \to 0).$$

Thus $\lambda = \mu$, $h + i\lambda \in \mathrm{Her}(A)$, and since $\mathrm{Sp}(h^*) = \mathrm{Sp}(h)^*$ and also $\mathrm{Sp}(h + i\lambda) \subset \mathbb{R}$, it follows that $\lambda = 0$, $h \in \mathrm{Her}(A)$. The Vidav-Palmer theorem has useful applications even in the commutative case.

EXAMPLE 9: Let E be a compact Hausdorff space and let $A = C(E)$, the algebra of all continuous complex functions on E. Let $\|\ \|$ be a Banach algebra norm on A such that $\|1\| = 1$ and $\|\,|a|\,\| = \|a\|$ for all $a \in A$. Then $\|\ \|$ is the supremum norm.

To see this, let $h \in A$ with $h(E) \subset \mathbb{R}$. Then for $t \in \mathbb{R}$, we have $\|\exp(ith)\| = \|\,|\exp ith|\,\| = \|1\| = 1$. Thus $h \in \mathrm{Her}(A)$ and so $J(A) = A$. Therefore, A is a B^*-algebra with the natural involution, and, since A is also commutative, we have $\|a\| = r(a) = \|a\|_\infty$.

R. T. Moore (1971) has obtained the following characterization of B^*-algebras which may be regarded as a "dual" of the Vidav-Palmer theorem.

THEOREM 15: *Let* Her(A') *denote the real linear span of* $D(1)$. *Then* $J(A) = A$ *if and only if* Her(A') $\cap$ i Her(A') = $\{0\}$.

8. INEQUALITIES AND EXTREMAL ALGEBRAS

For $a \in A$ we write $v(a)$ for the numerical radius of a; that is, $v(a) = \max |V(a)|$. In Theorem 12 we applied a contour integration argument to establish that $\|a\| \leq ev(a)$. Given an element a of A such that $V(a) \subset \mathbb{E}$, where $\mathbb{E}$ denotes the closed unit disc in $\mathbb{C}$, the norm of a is at most e. It is natural to seek estimates for the norms of polynomials in a.

THEOREM 16: *If* $a \in A$ *and* $n = 1, 2, 3, \ldots$, *then*

$$\|a^n\| \leq n! \left(\frac{e}{n}\right)^n v(a)^n. \tag{8.1}$$

Proof: Let $v(a) \leq 1$ so that, by (6.3), we have $\|\exp(za)\| \leq e^{|z|}$ for $z \in \mathbb{C}$. Let Γ_n denote the circle $\{z : |z| = n\}$ with positive orientation. Since

$$a^n = \frac{n!}{2\pi i} \int_{\Gamma_n} \exp(za) \frac{dz}{z^{n+1}},$$

it follows that $\|a^n\| \leq n! e^n / n^n$, as required.

Recall from Stirling's formula that $n!(e/n)^n \leq e\sqrt{n}$, and, asymptotically, $n!(e/n)^n \sim \sqrt{2\pi n}$. The estimate in Theorem 16 (which is in fact best possible) thus contrasts with the special case of B^*-algebras, where the power inequality gives $\|a^n\| \leq 2v(a)^n$ for $n = 1, 2, 3, \ldots$.

In seeking to establish that Theorem 16 is best possible we need to find the largest possible norm on a given polynomial in a subject to the constraint that $V(a) \subset \mathbb{E}$. More generally let K be a compact convex subset of $\mathbb{C}$ with at least two points. It is of interest to consider the constraint that $V(a) \subset K$. (For example, if $K = [-1, 1]$, we are seeking the largest norm on $p(h)$ when h is Hermitian with numerical radius 1.) Let Θ be the algebra of complex polynomials in a single indeterminate u; to be specific we might represent u by the complex function $u(z) = z$ for $z \in K$. Let $\{\|\ \|_\lambda : \lambda \in \Lambda\}$ be the collection of all algebra norms on Θ for which $V(u) \subset K$, and for $p \in \Theta$ let

$$\|p(u)\|_e = \sup\{\|p(u)\|_\lambda : \lambda \in \Lambda\}.$$

For each $\lambda \in \Lambda$ we have $\|u\|_\lambda \leq e \max|K|$, and it follows that $\|p(u)\|_e$ is finite for each $p \in \Theta$. Evidently $\|\ \|_e$ is an algebra norm on Θ; the completion of Θ with respect to $\|\ \|_e$ will be denoted by $\mathrm{Ea}(K)$. We readily verify, using (6.3), that $V(u) \subset K$ for the norm $\|\ \|_e$. For obvious reasons $\mathrm{Ea}(K)$ is called the *extremal algebra* for K.

The above construction of $\mathrm{Ea}(K)$, which is essentially due to P. G. Dixon [7], does not enable us to evaluate $\|p(u)\|_e$. By substantial use of function theory we may identify the dual space of $\mathrm{Ea}(K)$ with the Banach space $D(K)$ of all entire functions ϕ such that

$$\|\phi\| = \sup\left\{\frac{|\phi(\zeta)|}{\omega(\zeta)} : \zeta \in \mathbb{C}\right\} < \infty,$$

where $\omega(\zeta) = \max\{|\exp(\zeta z)| : z \in K\}$. Under this identification we obtain, in particular,

$$\|u^n\|_e = \sup\{|\phi^{(n)}(0)| : \phi \in D(K), \|\phi\| = 1\}, \tag{8.2}$$

$$V(u^n) = \{\phi^{(n)}(0) : \phi \in D(K), \|\phi\| = 1 = \phi(0)\}. \tag{8.3}$$

(For details see, for example, NR II, §24.) It is now easy to show that Theorem 16 is best possible.

EXAMPLE 10: In the algebra $\mathrm{Ea}(\mathbb{E})$, where $\mathbb{E}$ is the closed unit disc, we have $\|u^n\| = n!(e/n)^n$ for $n = 1, 2, 3, \ldots$.

For in this case, $\omega(\zeta) = e^{|\zeta|}$, and for $n = 1, 2, 3, \ldots$, we let $\phi(\zeta) = (e\zeta/n)^n$ ($\zeta \in \mathbb{C}$). Then $\phi \in D(\mathbb{E})$, and $\phi^{(n)}(0) = n!(e/n)^n$. By elementary calculus, $\|\phi\| = \sup_{t>0}(e/n)^n t^n e^{-t} = 1$. Therefore $\|u^n\|_e \geq n!(e/n)^n$ by (8.2), and the opposite inequality is given by (8.1). (As a corollary we note the failure of the inequality $v(u^n) \leq v(u)^n$ in the algebra $\mathrm{Ea}(\mathbb{E})$.)

The algebra Ea($[-1,1]$) gives interesting results about Hermitian elements. For this case, we have $\omega(\zeta) = e^{|\mathrm{Re}\,\zeta|}$ and so $\|u\|_e = 1$ by (8.2) and the classical Bernšteĭn theorem. It follows that $v(h) = \|h\|$ for any Hermitian element h, and the elementary fact that $r(h) = v(h)$ (see Theorem 13) combines to give a proof of Sinclair's theorem that $r(h) = \|h\|$.

We can also see that u^2 is very far from being Hermitian in Ea($[-1,1]$). Given $|\alpha| \leq 1$, let $\phi(\zeta) = \frac{1}{2}(1-\alpha) + \frac{1}{2}(1+\alpha)\cosh\zeta$ for $\zeta \in \mathbb{C}$. The Phragmén-Lindelöf theorem shows that $\phi \in D([-1,1])$, $\|\phi\| = 1 = \phi(0)$, and since $\phi''(0) = \frac{1}{2}(1+\alpha)$, relation (8.3) gives that $\{z : |z - \frac{1}{2}| \leq \frac{1}{2}\} \subset V(u^2)$. More is known about $V(u^2)$, but the explicit determination of the boundary of $V(u^2)$ remains unresolved —a problem which might have appealed to Cauchy!

There are many other extremal problems in numerical ranges that involve similar interplay with function theory. We consider the case in which the numerical range of each element is constrained to be as small as possible, namely the convex hull of the spectrum.

THEOREM 17: *Let $V(a) = \mathrm{co}\,\mathrm{Sp}(a)$ for all $a \in A$. Then we have $\|a^n\| \leq \frac{1}{2}ev(a)^n$ for $n = 1, 2, 3, \ldots$.*

Proof: Since $V(a) = \mathrm{co}\,\mathrm{Sp}(a)$ it follows from (6.3) and (6.5) that $\|\exp(a)\| = r(\exp(a))$. Let $r(a) < 1$ and let Γ denote the circle $\{z : |z| = 1\}$ with positive orientation. Then

$$-\frac{2}{e}a = \frac{1}{2\pi i}\int_\Gamma \exp\bigl((za+1)(za-1)^{-1}\bigr)\frac{dz}{z^2}.$$

Given $z \in \Gamma$ and $\lambda \in \mathrm{Sp}((za+1)(za-1)^{-1})$, the functional calculus tells us that $\lambda = (z\alpha+1)(z\alpha-1)^{-1}$ with $\alpha \in \mathrm{Sp}(a)$, and since $r(a) < 1$ we therefore have $\mathrm{Re}\,\lambda < 0$, $|e^\lambda| < 1$. Therefore

$$\|\exp\bigl((za+1)(za-1)^{-1}\bigr)\| \leq 1 \quad \text{for } z \in \Gamma,$$

and so $(2/e)\|a\| \leq 1$. This gives $\|a\| \leq \frac{1}{2}ev(a)$ for each $a \in A$. Finally

$$\|a^n\| \leq \tfrac{1}{2}ev(a^n) = \tfrac{1}{2}er(a^n) = \tfrac{1}{2}er(a)^n = \tfrac{1}{2}ev(a)^n.$$

In fact, the condition in Theorem 17 is so restrictive that A must be commutative; for if $a, b \in A$ and $f \in A'$, then $F(z) = f(\exp(-za)b\exp(za))$ for $z \in \mathbb{C}$, gives a bounded entire function such that $0 = F'(0) = f(ba - ab)$ for all $f \in A'$. (For the special case $A = C(E)$, E a compact Hausdorff space, we even have $\|a\| = \|a\|_\infty$, by the Vidav-Palmer theorem.)

On the other hand, let A be the disc algebra with norm $\| \ \|_e$, defined by

$$\|a\|_e = \inf\{\Sigma |\alpha_n| \|b_n\|_\infty : b_n \in \exp A, a = \Sigma \alpha_n b_n\}$$

where $a = \Sigma \alpha_n b_n$ means that the series converges uniformly on the closed unit disc. It may be verified that $V(a) = \operatorname{co} \operatorname{Sp}(a)$ for $a \in A$, and an argument involving Schwarz's Lemma shows that $\|u\| = \frac{1}{2}e$, where $u(z) = z$. Thus Theorem 17 is best possible.

We conclude this section in a different vein with the following elementary exercise on idempotent elements.

EXAMPLE 11: Let $j^2 = j$ and let $v(j) = 1$; then $\|j\| = 1$.

Indeed, the inequality $\|\exp(tj)\| \leq e^t$ $(t > 0)$ gives $\|1 + (e^t - 1)j\| \leq e^t$, whence $\|j\| \leq (e^t + 1)/(e^t - 1)$ for $t > 0$, and so $\|j\| \leq 1$.

9. THE ESSENTIAL NUMERICAL RANGE

We have already observed that the algebra numerical range $V(a)$ is determined by the values of the norm on the linear span of 1 and a; in fact,

$$V(a) = \bigcap_{z \in \mathbb{C}} \{\lambda : |z - \lambda| \leq \|z - a\|\}. \tag{9.1}$$

It follows that the numerical range of a is unchanged if we work in any subalgebra of A that contains 1 and a. The situation for quotient algebras is rather different.

LEMMA 5: *Let K be a closed bi-ideal of A and let π be the canonical mapping of A onto the quotient algebra A/K. Then*

$$V(\pi(a)) = \bigcap \{V(a+k) : k \in K\}.$$

Proof: Let $z, \lambda \in \mathbb{C}$. By the definition of the norm in A/K, we have $|z-\lambda| \leq \|\pi(z-a)\|$ if and only if $|z-\lambda| \leq \|z-(a+k)\|$ for all $k \in K$. The lemma now follows from (9.1).

Now let X be a complex Banach space, let $A = BL(X)$ and take $K = KL(X)$, the set of all compact operators on X. It is well known that $KL(X)$ is a closed bi-ideal of $BL(X)$; moreover, the quotient algebra $BL(X)/KL(X)$ is called the *Calkin algebra over* X.

DEFINITION: The *essential numerical range* $V_e(T)$ is defined, for $T \in A$, by

$$V_e(T) = V(\pi(T)),$$

where $V(\pi(T))$ is the algebra numerical range in the Calkin algebra over X.

Thus $V_e(T)$ is a non-empty compact convex subset of $\mathbb{C}$, and $V_e(T) = \{0\}$ if and only if T is compact. Lemma 5 gives

$$V_e(T) = \bigcap \{V(T+C) : C \in K\} \qquad (9.2)$$

where, in (9.2), $V(T+C)$ denotes the algebra numerical range of $T+C$ in A. It is also easy to show that

$$V_e(T) = \{f(T) : f \in D(I) \text{ and } f(K) = \{0\}\}.$$

DEFINITION: The *essential spectrum* $\mathrm{Sp}_e(T)$ is defined, for $T \in A$, by $\mathrm{Sp}_e(T) = \mathrm{Sp}(\pi(T))$.

We recall that T is a *Fredholm operator* if the kernel of T has finite dimension and the range of T is closed with finite codimension. Also T is Fredholm if and only if $\pi(T)$ is invertible (see [2]), and so $\lambda - T$ is Fredholm if and only if $\lambda \notin \mathrm{Sp}_e(T)$. It may be shown that each point of $\mathrm{Sp}(T) \setminus V_e(T)$ is an eigenvalue of T with finite dimensional eigenspace.

Given $T \in \mathrm{Her}(A)$ it is evident that $V_e(T) \subset \mathbb{R}$. Suppose conversely that $V_e(T) \subset \mathbb{R}$. We do not know in general if there exists $Q \in \mathrm{Her}(A)$ and $C \in K$ such that $T = Q + C$. That this is the case when X is a Hilbert space follows easily from the fact that the Hermitian elements of a B^*-algebra are precisely the self-adjoint elements.

The Calkin algebra over an arbitrary Banach space has been the object of considerable study (see [2]), but in this generality many questions about essential numerical ranges remain unresolved.

For the remainder of this section H denotes an infinite dimensional separable Hilbert space and $A = BL(H)$. We shall now denote the essential numerical range of $T \in A$ by $W_e(T)$ in order to relate it to the classical numerical range $W(T)$. Given a closed linear subspace M of H, we write P_M for the (orthogonal) projection of H onto M, and $\text{pr}_M(T)$ for the compression of T to M; that is,

$$\text{pr}_M(T) = P_M T | M.$$

The following theorem (1971) due to J. Anderson, P. A. Fillmore, J. G. Stampfli, and J. P. Williams is the starting point for many applications.

THEOREM 18: *For $T \in A$ the following statements are equivalent.*

(i) $\lambda \in W_e(T)$.

(ii) *There exists an infinite dimensional closed subspace M of H such that, relative to an orthonormal basis for M, $\text{pr}_M(T)$ has diagonal matrix representation with entries λ_n and $\lambda = \lim_{n \to \infty} \lambda_n$.*

(iii) *There exists a projection P of infinite rank such that $P(T - \lambda I)P \in K$.*

(iv) *There exists an orthonormal sequence $\{e_n\}$ such that $\lambda = \lim_{n \to \infty} \langle Te_n, e_n \rangle$.*

(v) *There exists a sequence $\{x_n\}$ in $S(H)$ that converges weakly to 0 and satisfies $\lambda = \lim_{n \to \infty} \langle Tx_n, x_n \rangle$.*

J. S. Lancaster [9] has recently used Theorem 18(v) to give a deep analysis of the boundary of $W(T)$, as follows.

THEOREM 19: *For $T \in A$ we have*

$$\text{ext cl}\, W(T) \subset W_e(T) \cup W(T), \; \text{cl}\, W(T) = \text{co}\{W_e(T) \cup W(T)\}.$$

Proof: Note that the second statement is an immediate consequence of the first and the fact that $W_e(T) \cup W(T) \subset \text{cl}\, W(T)$. We may suppose without loss of generality that $\text{cl}\, W(T)$ is contained in the closed right half-plane and that $0 \in \text{ext}\,\text{cl}\, W(T)$, $0 \notin W(T)$. We must thus show that $0 \in W_e(T)$.

By hypothesis there exists a sequence $\{x_n\}$ in $S(H)$ such that $\lim_{n\to\infty} \langle Tx_n, x_n \rangle = 0$. By the weak sequential compactness of the unit ball of H, we may suppose that $\{x_n\}$ converges weakly to x with $\|x\| \leq 1$. By Theorem 18(v), it is now enough to show that $x = 0$. We note that the identity

$$\|x - x_n\|^2 = \|x_n\|^2 - \|x\|^2 + 2\,\text{Re}\langle x - x_n, x\rangle$$

gives

$$\lim_{n\to\infty} \|x - x_n\|^2 = 1 - \|x\|^2. \tag{9.3}$$

Suppose that $\|x\| = 1$. Then, by (9.3), $\lim_{n\to\infty}\|x - x_n\| = 0$ and so

$$|\langle Tx, x\rangle| \leq |\langle T(x - x_n), x\rangle| + |\langle Tx_n, x - x_n\rangle| + |\langle Tx_n, x_n\rangle|$$

$$\leq |\langle x - x_n, T^*x\rangle| + \|T\|\,\|x - x_n\| + |\langle Tx_n, x_n\rangle|$$

$$\to 0 \quad \text{as} \quad n\to\infty.$$

This is a contradiction since $0 \notin W(T)$.

Suppose now that $0 < \|x\| < 1$. The main step of the proof of Theorem 1 gives $\lim_{n\to\infty}\|Sx_n\| = 0$, where $S = T + T^*$. Since

$$\langle Sx, x\rangle = \langle x - x_n, S^*x\rangle + \langle Sx_n, x\rangle$$

it follows that $\langle Sx, x\rangle = 0$, $\text{Re}\langle Tx, x\rangle = 0$. We have $\langle Tx, x\rangle \neq 0$ since $0 \notin W(T)$. Let $y_n = \|x - x_n\|^{-1}(x - x_n)$. Since

$$\langle T(x - x_n), x - x_n\rangle = \langle Tx, x - x_n\rangle - \langle Tx_n, x\rangle + \langle Tx_n, x_n\rangle$$

$$\to -\langle Tx, x\rangle \quad \text{as}\, n\to\infty,$$

it follows from (9.3) that

$$\lim_{n\to\infty} \langle Ty_n, y_n\rangle = -\bigl(1 - \|x\|^2\bigr)^{-1}\langle Tx, x\rangle.$$

We have now shown that there exist $\alpha, \beta > 0$ with $i\alpha, -i\beta \in \text{cl } W(T)$, contrary to the hypothesis that $0 \in \text{ext cl } W(T)$. The proof is complete.

There are numerous corollaries of Lancaster's theorem. For example, $W(T)$ is closed if and only if $W_e(T) \subset W(T)$. In particular, if T is compact, then $W(T)$ is closed if and only if $0 \in W(T)$. It may also be shown that if T is compact, then $W(T)$ cannot be a closed semidisc (though it can be a closed disc).

Renewed interest in studying $W(T)$ in the 1950's was partly stimulated by problems about commutators; it turned out subsequently that the key to the situation lay rather with $W_e(T)$. To be precise, Theorem 18(ii) is the starting point for the proof of the following intriguing theorem of J. Anderson (1971).

THEOREM 20: *Let $T \in A$. Then $0 \in W_e(T)$ if and only if we have $T = RS - SR$ for some $R, S \in A$ with $R^* = R$.*

Together with some arguments of J. P. Williams, Anderson's theorem leads to the following sharp version of the Brown-Pearcy theorem on commutators.

THEOREM 21: *Let $T \in A$. Then $T = RS - SR$ for some $R, S \in A$ if and only if, for every non-zero λ, $T - \lambda I \notin K$. If the condition is satisfied, then R can be chosen to be similar to a self-adjoint operator.*

Finally, we refer the reader to NR II for other generalizations and applications of numerical ranges.

REFERENCES

Books

NR I F. F. Bonsall and J. Duncan, *Numerical Ranges of Operators on Normed Spaces and of Elements of Normed Algebras*, London Math. Soc. Lecture Note Series, no. 2, Cambridge University Press, Cambridge-New York, 1971.

NR II _____, *Numerical Ranges II*, London Math. Soc. Lecture Note Series, no. 10, Cambridge University Press, Cambridge-New York, 1973.

1. _____, *Complete Normed Algebras*, Springer-Verlag, New York, 1973.
2. S. R. Caradus, W. E. Pfaffenberger and B. Yood, *Calkin Algebras and Algebras of Operators on Banach Spaces*, Marcel Dekker, New York, 1974.
3. P. A. Fillmore, *Notes on Operator Theory*, D. Van Nostrand, Princeton, N.J., 1970.
4. P. R. Halmos, *Hilbert Space Problem Book*, D. Van Nostrand, Princeton, N.J., 1967.

Articles

5. B. Bollobás, "The spectral decomposition of compact Hermitian operators on Banach spaces," *Bull. London Math. Soc.*, 5 (1973), 29–36.
6. M. J. Crabb, "The numerical range of an unbounded operator," *Proc. Amer. Math. Soc.*, 55 (1976), 95–96.
7. P. G. Dixon, "Varieties of Banach algebras," *Quart. J. Math. Oxford, Ser.* (2) 27 (1976), 481–487.
8. P. R. Halmos, *Math. Reviews*, 41 (1971), no. 1368.
9. J. S. Lancaster, "The boundary of the numerical range," *Proc. Amer. Math. Soc.*, 49 (1975), 393–398.
10. B. Schmidt, "Über die Ecken des numerischen Wertebereichs in einer Banachalgebra," *Math. Z.*, 126 (1972), 47–50.
11. J. G. Stampfli, "Extreme points of the numerical range of a hyponormal operator," *Michigan Math. J.*, 13 (1966), 87–89.
12. A. Wintner, "Zur Theorie der beschränkten Bilinearformen," *Math. Z.*, 30 (1929), 228–282.

PROJECTION OPERATORS IN APPROXIMATION THEORY

E. W. Cheney

1. A RECONNAISSANCE OF APPROXIMATION THEORY

Approximation Theory traditionally concerns itself with the approximation of functions, primarily functions of a real or complex variable. As we shall see, some newer aspects of the subject concern the approximation of other objects, particularly operators.

A typical situation within the purview of approximation theory is the following. One is presented with a metric space (X,d) whose elements are the objects to be approximated. Within X a subset Y is prescribed. The elements of Y are the "approximators." For any $x \in X$ and any $y \in Y$, $d(x, y)$ is interpreted as the error or discrepancy when y is used as an approximation to x. The quantity $\text{dist}(x, Y) = \inf_{y \in Y} d(x, y)$ measures how well x is capable of being approximated by elements of Y. For a fixed pair $Y \subset X$, a theory of approximation would attempt to answer such questions as the following:

(1) Does there exist, for an arbitrary x in X, a "best" approximation y in Y, i.e., an element for which $d(x,y) = \text{dist}(x, Y)$?

(2) Is it possible to estimate dist(x, Y) from a knowledge of special properties of x (as, for example, its differentiability if it is a function)?

(3) Does there exist a continuous "proximity map" $A: X \to Y$, i.e., a map such that $d(x, Ax) = \text{dist}(x, Y)$ for all $x \in X$?

(4) Is it true that each x has a *unique* best approximation in Y?

(5) For a fixed x in X, how are its best approximations in Y characterized?

(6) What algorithms (numerical procedures) can be devised for producing best approximations?

(7) Do there exist simple and convenient substitutes for the proximity map? In particular, do such maps $B: X \to Y$ exist satisfying $d(x, Bx) \leq (1 + \epsilon) \text{dist}(x, Y)$, with ϵ small?

The above questions concern a fixed pair, $Y \subset X$. If there is a nested sequence of sets, $Y_1 \subset Y_2 \subset \cdots \subset X$, then some questions about asymptotics can be raised:

(8) What is the nature of the sequence $e_n(x) = \text{dist}(x, Y_n)$? In particular, does it converge to 0?

(9) Can the rapidity of convergence of the sequence be related to the structural properties of the element x?

Another branch of approximation theory concerns "optimal" choices for an approximating subspace Y. For example, if K is a set and if Y is a finite dimensional subspace in a normed (real) linear space X, then the number

$$E(K, Y) = \sup_{x \in K} \text{dist}(x, Y)$$

measures how well Y serves as an approximating class for the elements of K. Then one can seek a subspace Y which minimizes $E(K, Y)$ under the constraint that the dimension of Y does not exceed n. Indeed, the *n-width* of K is defined to be

$$\delta_n(K) = \inf \{ E(K, Y) : Y \subset X, \dim Y \leq n \}.$$

In order to illustrate the manner in which the above questions might be answered in a specific situation, let us cite some theorems from classical approximation theory concerning the space $X = C[-1, 1]$ and the subspace $Y = \Pi_{n-1}$ of polynomials of degree $< n$.

The space X consists of all real-valued continuous functions on the interval $[-1,1]$ with norm $\|x\| = \max_{-1 \leq t \leq 1} |x(t)|$. The metric on X is defined by $d(x,z) = \|x - z\|$.

THEOREM 1: *Each element x of $C[-1,1]$ possesses a unique best approximation y in Π_{n-1}. The mapping $x \mapsto y$ is continuous. The element y is characterized by the existence of $n+1$ points t_i satisfying $-1 \leq t_0 < \cdots < t_n \leq 1$, such that*

$$y(t_i) = x(t_i) + (-1)^i \epsilon, \quad \text{where } |\epsilon| = \|x - y\|.$$

THEOREM 2: *For each $x \in C[-1,1]$ the sequence $e_n(x) = \operatorname{dist}(x, \Pi_n)$ converges to 0 as $n \to \infty$. If x has a continuous pth derivative, then $e_n(x)$ converges to 0 at least as rapidly as the sequence n^{-p}. Conversely, if $e_n(x) = O(n^{-p-\epsilon})$ for some $\epsilon > 0$, then x possesses a continuous pth derivative on the open interval $(-1,1)$.*

THEOREM 3: *If $x \in C[-1,1]$ and if Lx is the unique element of Π_n that takes the same value as x at the points $\cos(i\pi/n)$ ($0 \leq i \leq n$), then L is a linear projection operator and satisfies*

$$\|x - Lx\| \leq (2 + (2/\pi)\log n)\operatorname{dist}(x, \Pi_n).$$

THEOREM 4: *For a fixed integer p, let K denote the class of all functions x in $C[-1,1]$ whose pth derivative exists a.e. and satisfies $|x^{(p)}(t)| \leq 1$ a.e. Then $E(K, \Pi_{n-1})$ is asymptotic to $c_p n^{-p}$ as $n \to \infty$. The n-width of K is asymptotic to $(2/\pi)^p c_p n^{-p}$. Thus algebraic polynomials are not optimal for approximating the elements of K.*

2. LINEAR APPROXIMATION OPERATORS

Consider a pair of normed (real) linear spaces X, Y, with $Y \subset X$ and $\dim(Y) < \infty$. As in the previous section, there is a proximity map $A: X \to Y$ such that $\|x - Ax\| = \operatorname{dist}(x, Y)$ for each $x \in X$. If X is a Hilbert space, then A is the familiar orthogonal projection, which is linear and of norm 1. In most situations, however, A is nonlinear and difficult to use. It is therefore important to find simpler maps, linear if possible, which can be useful in providing approximations.

For example, if $X = C[0,1]$ and $Y = \Pi_n$, then the map $B_n: X \to Y$ defined by

$$B_n x = \sum_{i=0}^{n} x\left(\frac{i}{n}\right) y_{ni}, \quad \text{where} \quad y_{ni}(t) = \binom{n}{i} t^i (1-t)^{n-i},$$

produces a polynomial approximation to x which satisfies

$$\|x - B_n x\| \leq \frac{5}{4} \omega(x; n^{-1/2}). \tag{1}$$

In this inequality, ω is the modulus of continuity, defined by

$$\omega(x; \delta) = \sup\{|x(t) - x(s)| : 0 \leq t \leq s \leq t + \delta \leq 1\}.$$

Since continuity on $[0,1]$ implies *uniform* continuity, we have $\lim_{\delta \to 0} \omega(x; \delta) = 0$ for each $x \in C[0,1]$. Consequently, $\|x - B_n x\| \to 0$ for each x as $n \to \infty$. The linear operators B_n, called the *Bernstein* [Bernšteĭn] *operators*, therefore have the ability to produce approximations of any desired precision, and do so in a simple and elegant manner. The theory of these operators is in a high state of refinement, and the interested reader should refer to Lorentz's monograph [34]. The best constant in the inequality (1) was determined by Sikkema [48].

The quality of the approximations produced by the operators B_n leaves much to be desired, however. Thus for the elementary function $x(t) = t^2$ we find that $\|x - B_n x\| = (4n)^{-1}$ for $n \geq 2$, and indeed this is typical. A precise result, known as a "saturation theorem," was given by Bajšanski and Bojanić [1]: If $x \in C[0,1]$ and if $\|x - B_n x\| = o(n^{-1})$ as $n \to \infty$, then $x(t) = at + b$ for some a and b. It appears, therefore, that the operators B_n are not effective in producing approximations comparable in accuracy to the best approximations in this setting; for, as we noted in Theorem 2, $\text{dist}(x, \Pi_n) = O(n^{-p})$ for any p-times continuously differentiable function x.

If the search for good approximation operators is confined to the class of continuous linear maps, $L: X \to Y$, then one advantage to be gained is that such maps have a simple form. Namely, if $\{y_1,\ldots,y_n\}$ is any basis for Y, then L is expressible in the form

$$Lx = \sum_{i=1}^{n} f_i(x) y_i, \qquad (2)$$

where the functionals f_i are real-valued and defined on X. Elementary arguments show that each f_i is linear and continuous. Thus f_i belongs to the conjugate space X^*. Equation (2) can also be written in the equivalent tensor form

$$L = \sum_{i=1}^{n} f_i \otimes y_i. \qquad (3)$$

If no additional assumptions are made, almost nothing useful can be said about the accuracy of the approximations produced by L (i.e., about $\|x - Lx\|$). The missing ingredient here is an assumption about how close the numbers $\|x - Lx\|$ and dist(x, Y) are to each other. Let us impose, then, an eminently reasonable condition

$$\|x - Lx\| \leq k \, \text{dist}(x, Y) \qquad (x \in X) \qquad (4)$$

where k is a constant independent of x.

For example, suppose that $L: C[a,b] \to \Pi_n$ and that $\|x - Lx\| \leq 10\,\text{dist}(x, \Pi_n)$. Then for each x, the element Lx is an approximation to x that forfeits only one decimal place of accuracy in comparison with the best approximation. In practical terms, this may be a small price to pay for other favorable properties that L may possess, such as simplicity (and, of course, linearity).

If L is a linear map from X to Y satisfying (4), then it follows that $Ly = y$ for each $y \in Y$. A bounded linear map $L: X \to Y$ having this property is called a *projection* of X onto Y. Most of the linear maps which are useful in providing good approximations are projections, and classical analysis abounds in examples of them. Even if we did not insist on the property expressed in (4), it is still natural to require of an approximation operator $L: X \to Y$ that it leave unmolested the elements of Y. This seemingly weaker assumption leads to (4), however.

THEOREM 5: *If P is a projection of the normed space X onto its subspace Y, then*

$$\|x - Px\| \leq \|I - P\| \operatorname{dist}(x, Y) \quad \text{for all } x \in X.$$

Proof: For any $y \in Y$, we have $\|x - Px\| = \|(x-y) - P(x-y)\|$ $= \|(I-P)(x-y)\| \leq \|I-P\| \|x-y\|$. Now take the infimum as y ranges over Y. □

Now let us cite some important examples, to which we shall refer later.

EXAMPLE 1 (Lagrange Interpolation): Let $X = C[a,b]$ and let $Y = \Pi_{n-1}$. Select arbitrary distinct points $t_1, \ldots, t_n$ in the interval $[a,b]$, and define

$$Px = \sum_{i=1}^{n} x(t_i) y_i, \quad \text{where} \quad y_i(t) = \prod_{\substack{j=1 \\ j \neq i}}^{n} \frac{t - t_j}{t_i - t_j} \quad \text{for } i,j = 1, \ldots, n.$$

One can see easily that $y_i \in Y$ and $y_i(t_j) = \delta_{ij}$. It follows that Px is a polynomial of degree at most $n-1$ and agrees with x at the points $t_1, \ldots, t_n$. This in turn implies that if $y \in Y$ then $Py = y$ (for $Py - y$ is a polynomial of degree $\leq n-1$ having n roots, and must therefore be 0). The projection P can be written in tensor form as

$$P = \sum_{i=1}^{n} \hat{t}_i \otimes y_i,$$

where the notation $\hat{t}$ denotes the point-functional corresponding to t: $\hat{t}(x) = x(t)$ for all $x \in C[a,b]$.

EXAMPLE 2: This projection is obtained by the "least-squares theory." Again let $X = C[a,b]$ and $Y = \Pi_{n-1}$; let w be a positive and continuous function on $[a,b]$. An inner product can be introduced in X by the equation

$$\langle u, v \rangle = \int_a^b u(t) v(t) w(t) \, dt.$$

Let $\{y_1, \ldots, y_n\}$ be a basis for Y which is orthonormal with respect to this inner product; thus $\langle y_i, y_j \rangle = \delta_{ij}$. Then a projection can be defined by the equation

$$Px = \sum_{i=1}^{n} \langle x, y_i \rangle y_i.$$

This can be termed the *orthogonal projection* (determined by w) from $C[a,b]$ onto Π_{n-1}.

3. PROJECTION CONSTANTS

In the preceding section, we saw that if P is a projection of a normed space X onto a subspace Y of X, then (since $\|I - P\| \leq 1 + \|P\|$) it follows that

$$\|x - Px\| \leq (1 + \|P\|)\text{dist}(x, Y).$$

Thus, there are sound practical reasons for seeking those projections of X onto Y for which $\|P\|$ is least. The number

$$\lambda(Y, X) = \inf\{\|P\| : P \text{ is a projection of } X \text{ onto } Y\}$$

is known as the *relative projection constant* of Y in X. If there exist no projections of X onto Y we write $\lambda(Y, X) = \infty$. If $\lambda(Y, X) < \infty$, then Y is said to be *complemented* in X. The number

$$\lambda(Y) = \sup\{\lambda(Y, X) : X \supset Y\}$$

is the *absolute projection constant* of Y.

THEOREM 6 (cf. [22]): *The infimum defining the relative projection constant $\lambda(Y, X)$ is attained if Y is finite dimensional.*

Proof (outline): We shall prove somewhat more; namely, the conclusion is true if Y is complemented in X and is the conjugate of another space, $Y = Z^*$. In the space $\mathcal{L}(X, Y)$ of all bounded linear maps from X into Y there is a topology known as the weak*-operator topology, w*o. A net L_α converges to 0 in the w*o-topology if and only if $(L_\alpha x, z) \to 0$ for all $x \in X$ and $z \in Z$. (Here $(f, z) = f(z)$ for $z \in Z$ and $f \in Z^*$.) Since Y is complemented in X, there is a projection P_0 of X onto Y. We only need consider the set

$$\{P: P \text{ is a projection of } X \text{ onto } Y \text{ and } \|P\| \leq \|P_0\|\}.$$

This set is w*o-compact, and $\|P\|$ is w*o-lower-semicontinuous. Hence the infimum is attained. □

THEOREM 7 (cf. [26]): *Each n-dimensional Banach space X satisfies $\lambda(X) \leq \sqrt{n}$.*

Proof: It is a consequence of a lemma of F. John [23] that there exist an isomorphism $L: X \to l_2^{(n)}$ and a subset $\{v_1, \ldots, v_s\}$ of $l_2^{(n)}$ such that $\|L^{-1}\| \leq 1$, $\sum_{i=1}^{s} \|L^* v_i\|^2 \leq n$, and $u = \sum_{i=1}^{s} (u, v_i) v_i$ for all $u \in l_2^{(n)}$. It follows at once that $\|u\|^2 = (u, u) = \sum_{i=1}^{s} (u, v_i)^2$.

Suppose now that $X \subset Z$. By the Hahn-Banach Theorem, each functional $L^* v_i$ has a norm-preserving extension f_i in Z^*. Define $P = \sum_{i=1}^{s} f_i \otimes L^{-1} v_i$; clearly P maps Z into X. If $x \in X$, then

$$Px = \sum f_i(x) L^{-1} v_i = \sum (L^* v_i, x) L^{-1} v_i$$
$$= L^{-1} \sum (v_i, Lx) v_i = L^{-1} \sum (Lx, v_i) v_i = L^{-1} Lx = x.$$

Thus P is a projection of Z onto X. In order to estimate $\|P\|$, let $z \in Z$, $\|z\| \leq 1$, $\phi \in X^*$, and $\|\phi\| \leq 1$. By the Cauchy-Schwarz Inequality,

$$\phi Pz = \sum f_i(z) \phi L^{-1} v_i \leq \sum \|f_i\| |\phi L^{-1} v_i|$$
$$\leq \left(\sum \|f_i\|^2\right)^{1/2} \left(\sum |\phi L^{-1} v_i|^2\right)^{1/2}$$
$$\leq \left(\sum \|L^* v_i\|^2\right)^{1/2} \left(\sum (L^{*-1}\phi, v_i)^2\right)^{1/2}$$
$$\leq (\sqrt{n}) \|L^{*-1} \phi\| \leq \sqrt{n}.$$

Hence $\|P\| \leq \sqrt{n}$. □

The importance of projection constants extends far beyond the applications in approximation theory to which we have already alluded. In order to illustrate another application, consider a linear operator $L: Y \to Z$ and suppose that $Y \subset X$. One can ask whether L has an extension $\bar{L}: X \to Z$. If Z is the real line, the best possible answer is provided by the Hahn-Banach Theorem, but in the general case an extension need not exist. If, however, a projection P exists from X onto Y then LP is an extension of L. The following theorem describes this situation in greater detail, and connects it with the possibility of isometric imbeddings of Y into the space $l_\infty(S)$ of bounded functions on some set S.

THEOREM 8: *For a Banach space Y and a real number λ, $\lambda \geq 1$, the following properties are equivalent.*

(1) *For every Banach space X containing Y, there is a projection of norm λ from X onto Y.*

(2) *If $L \in \mathcal{L}(Y, Z)$ and $Y \subset X$, then L has an extension $L' \in \mathcal{L}(X, Z)$ satisfying $\|L'\| \leq \lambda \|L\|$.*

(3) *When Y is imbedded isometrically in a space $l_\infty(S)$, there is a projection of norm λ from $l_\infty(S)$ onto Y.*

(4) *If $L \in \mathcal{L}(Z, Y)$ and $Z \subset X$, then L has an extension $L' \in \mathcal{L}(X, Y)$ satisfying $\|L'\| \leq \lambda \|L\|$.*

Proof: (1) implies (2) because a suitable extension of L is LP, where P is a projection of norm λ from X onto Y.

(2) implies (3) as follows. In (2) let $X = l_\infty(S)$ and $Z = Y$. Let L be the identity map on Y. The extension L' is a projection with norm λ.

For the implication of (4) from (3), let P be a projection of norm λ from $l_\infty(S)$ onto Y. Let $L \in \mathcal{L}(Z, Y)$ and $Z \subset X$. If $s \in S$ and $z \in Z$, put $f_s(z) = (Lz)(s)$. Then $f_s \in Z^*$ and $\|f_s\| \leq \|L\|$. By the Hahn-Banach Theorem, f_s has a norm-preserving extension F_s in X^*. If $x \in X$ and $s \in S$, put $(L'x)(s) = F_s(x)$. Then L' is an extension of L satisfying $\|L'\| = \|L\|$. Hence PL' is an extension of L in $\mathcal{L}(X, Y)$ satisfying $\|PL'\| \leq \lambda \|L\|$.

That (4) implies (1) is obtained by taking $Z = Y$ and L to be the identity on Y in (4). □

COROLLARY 1: *Each space $l_\infty(\Gamma)$ has projection constant 1.*

COROLLARY 2: *If $Y \subset X$, then $\lambda(Y) \leq \lambda(X) \lambda(Y, X)$.*

Proof: If both factors in the right side are finite, it is clear that $\lambda(Y)<\infty$. Let P be any projection of X onto Y. Let $A \subset B$ and let $L \in \mathcal{L}(A,Y)$. Then $L \in \mathcal{L}(A,X)$. For any $\epsilon > 0$, by the equivalence of (1) and (4) in the preceding theorem, L has an extension $L' \in \mathcal{L}(B,X)$ such that $\|L'\| \leq [\epsilon + \lambda(X)]\|L\|$. Observe that L has PL' as an extension in $\mathcal{L}(B,Y)$ and that

$$\|PL'\| \leq \|P\|[\epsilon + \lambda(X)]\|L\|.$$

Hence we have $\lambda(Y) \leq \|P\|[\epsilon + \lambda(X)]$. Taking infima, we obtain $\lambda(Y) \leq \lambda(Y,X)\lambda(X)$. □

COROLLARY 3: *If L is an isomorphism between two normed spaces Y and Z, then $\lambda(Y) \leq \|L\|\|L^{-1}\|\lambda(Z)$.*

THEOREM 9: *If Y is a finite dimensional subspace of $X = C(T)$, where T is a compact Hausdorff space, then $\lambda(Y) = \lambda(Y,X)$. If T is extremally disconnected, then this equation is true even for subspaces of infinite dimension.*

Proof (H. E. Lacey): It is known [32, p. 232, Theorem 2] that for any $\mu > 1$ there is a subspace Z such that $Y \subset Z \subset C(T)$ and such that an isomorphism $L: Z \to l_m^\infty$ exists satisfying $\|L\|\|L^{-1}\| \leq \mu$. By Corollaries 1 and 3, $\lambda(Z) \leq \mu$. Now let P be any projection of $C(T)$ onto Y. Then the restricted operator $P|Z$ is a projection of Z onto Y. Hence by Corollary 2,

$$\lambda(Y) \leq \lambda(Z)\lambda(Y,Z) \leq \mu\|P|Z\| \leq \mu\|P\|.$$

By taking infima on μ and P, we obtain the desired inequality.

If T is extremally disconnected (i.e., the closure of each open set is open), then $\lambda(C(T)) = 1$. In fact these conditions are equivalent for a compact Hausdorff space T; see Day [8, p. 95] for the proof and references. Now by Corollary 2, any subspace Y satisfies $\lambda(Y) \leq \lambda(Y,C(T))$. □

The next theorem, giving an intrinsic characterization of $\lambda(X)$, was communicated privately by C. Franchetti.

THEOREM 10: *The projection constant of a finite dimensional space X satisfies the equation $\lambda(X) = \inf \max \|\Sigma \sigma_i x_i\|$, where the maximum is for all choices $\sigma_i = \pm 1$, and the infimum is over all finite sets $\{x_1, \ldots, x_s\} \subset X$ such that $\Sigma_{i=1}^s f_i \otimes x_i = I$ for some $f_i \in X^*$ satisfying $\|f_i\| \leq 1$.*

Proof: Consider the compact Hausdorff space $T = \{f \in X^* : \|f\| \leq 1\}$. For each $x \in X$ define $\hat{x} : T \to \mathbb{R}$ by $\hat{x}(f) = f(x)$. Let $\hat{X} = \{\hat{x} : x \in X\}$. The map $x \mapsto \hat{x}$ imbeds X isometrically as a subspace $\hat{X}$ of $C(T)$. By the preceding theorems, $\lambda(X) = \lambda(\hat{X}) = \lambda(\hat{X}, C(T))$.

Now suppose that $\Sigma_{i=1}^s f_i \otimes x_i = I$ and $\|f_i\| \leq 1$. Define $P : C(T) \to \hat{X}$ by putting $Pz = \Sigma z(f_i) \hat{x}_i$. The map P is a projection onto $\hat{X}$ because $P\hat{x} = \Sigma \hat{x}(f_i) x_i = \Sigma f_i(x) x_i = x$. Since $P = \Sigma \hat{f}_i \otimes \hat{x}_i$ ($\hat{f}_i =$ point functional on $C(T)$), we have $\lambda(X) \leq \|P\| = \|\Sigma |\hat{x}_i| \|_\infty = \sup_{f \in T} \Sigma |\hat{x}_i(f)| = \sup_{f \in T} \max_{\sigma_i = \pm 1} \Sigma \sigma_i \hat{x}_i(f) = \max_{\sigma_i} \sup_f f(\Sigma \sigma_i x_i) = \max_{\sigma_i} \|\Sigma \sigma_i x_i\|$. (See Section 5 below.)

In order to complete the proof we use the fact that for any $\epsilon > 0$ there is a projection P with finite carrier from $C(T)$ onto $\hat{X}$ satisfying $\|P\| < \lambda(\hat{X}) + \epsilon$. (See Section 5 below.) Write $Pz = \Sigma_{i=1}^s z(f_i) \hat{x}_i$, using appropriate $f_i \in T$ and $x_i \in X$. Then, as above, $\|P\| = \max_{\sigma_i} \|\Sigma_{i=1}^s \sigma_i x_i\| < \lambda(X) + \epsilon$. Now it is only necessary to observe that $\Sigma f_i \otimes x_i = I$, since for every $x \in X$, we have $\hat{x} = P\hat{x} = \Sigma \hat{x}(f_i) \hat{x}_i = \Sigma f_i(x) \hat{x}_i = [\Sigma f_i(x) x_i]\hat{}$ whence $x = \Sigma f_i(x) x_i$. $\square$

4. MINIMAL PROJECTIONS

If P is a projection of a normed space X onto a subspace Y of X and if $\|P\| = \lambda(Y, X)$, then we term P a *minimal projection*. Extending the concept slightly, we say that P is *minimal in a class of projections* if $\|P\| \leq \|Q\|$ for every Q in that class. For the existence question, we have the following refinement of Theorem 6.

THEOREM 11: *Let $Y \subset X$, and suppose that Y is (isometric to) a conjugate space $Y = Z^*$. If $\mathcal{Q}$ is any family of projections from X onto Y which is closed in the weak*-operator topology of $\mathcal{L}(X, Z^*)$, then $\mathcal{Q}$ contains a minimal element.*

Some related theorems occur in [21] and [37].
We now give some examples of minimal projections.

THEOREM 12: *The Fourier projection of $C_{2\pi}$ onto the nth order trigonometric polynomials is minimal.*

Proof: It is convenient to give the proof for the complex-valued function spaces. Put $e_k(t) = e^{ikt}$ and $\langle x, y \rangle = (1/2\pi)\int_0^{2\pi} x(t)\overline{y(t)}\, dt$. The subspace $°\Pi_n$ onto which we project has the orthogonal base $\{e_k : |k| \leq n\}$. The Fourier projection is $F_n x = \sum_{k=-n}^{n} \langle x, e_k \rangle e_k$. If P is any projection of $C_{2\pi}$ onto $°\Pi_n$, then the marvelous equation

$$F_n = \frac{1}{2\pi} \int_0^{2\pi} T_\lambda P T_{-\lambda}\, d\lambda \tag{1}$$

holds, in which T_λ denotes the shift operator defined by $(T_\lambda x)(t) = \hat{t}T_\lambda x = x(t+\lambda)$. The above equation is often referred to as the Berman-Marcinkiewicz equation because of Berman's work [3], and because an equation like this for interpolatory projections P was given by Marcinkiewicz [36]. The proof of (1) is effected by testing it on the functions e_k (whose linear combinations are dense in $C_{2\pi}$, by the Weierstrass Approximation Theorem). Equation (1) shows that F_n is the average of a family of projections $T_\lambda P T_{-\lambda}$ and it follows immediately that $\|F_n\| \leq \|P\|$. □

It is a standard result about Fourier series that F_n can be expressed in the form

$$(F_n x)(t) = \frac{1}{2\pi} \int_{-\pi}^{\pi} x(s+t) D_n(s)\, ds$$

in which $D_n(s) = \sin(n+\frac{1}{2})s / \sin\frac{1}{2}s$ (the Dirichlet kernel). Consequently, the norm of the operator F_n, which is obtained by taking the supremum of $\|F_n x\|$ as x ranges over the unit sphere of $C_{2\pi}$, is the number

$$\rho_n = \frac{1}{2\pi} \int_{-\pi}^{\pi} |D_n(s)|\, ds.$$

This number is known as the nth *Lebesgue constant*, after the work of Lebesgue [33]. Many other authors have studied this sequence, among them Szegö, Fejér, Watson, Rivlin, and Stečkin. Perhaps the most elegant result is this one of Galkin [13]:

$$\rho_n = (4/\pi^2)\log(2n+1) + 1 - \epsilon_n \qquad (0 < \epsilon_n < 0.0103).$$

The next example is a projection of rank n in the space $C[a,b]$. Select points ("nodes") $a = t_1 < t_2 < \cdots < t_n = b$. Let Y be the n-dimensional subspace of functions which are linear in each interval $[t_i, t_{i+1}]$. This is a useful subspace; it plays a rôle, for example, in the Trapezoid Rule for numerical integration. A natural projection onto Y is the operation of interpolation at the nodes. Thus for each $x \in C[a,b]$, Px is the unique element of Y which takes the same values at the nodes as x. For $t \in [t_i, t_{i+1}]$, $(Px)(t)$ is a convex combination of $x(t_i)$ and $x(t_{i+1})$. Hence $\|Px\| \leq \max_i |x(t_i)| \leq \|x\|$ and $\|P\| = 1$.

The subspace Y described above is obviously isometric to $l_\infty^{(n)}$, and it is therefore not surprising that it is the range of a norm-1 projection.

Next we consider a space Y in $C(T)$ which is isometric to $l_1^{(n)}$. Let T be the union of 2^n disjoint closed sets T_j, and let a basis $y_1, \ldots, y_n$ for Y be defined as follows: on T_j the function $y_i(t)$ is constantly equal to $(-1)^k$, where k is the greatest integer in $(j-1)2^{i-n}$. These functions are called *Rademacher functions* after the original work [43]. In order to define a projection, select a point $t_i \in T_i$ ($1 \leq i \leq 2^n$) and let $Px = \Sigma \langle x, y_i \rangle y_i$ with $\langle x, u \rangle = 2^{-n} \Sigma_{i=1}^{2^n} x(t_i) u(t_i)$.

The verification that this projection is minimal proceeds in four steps. First one proves that Y is isometric to $l_1^{(n)}$. In fact, for any n-tuple $(\alpha_1, \ldots, \alpha_n)$, we have $\|\Sigma_{i=1}^n \alpha_i y_i\|_\infty = \Sigma_{i=1}^n |\alpha_i|$. Secondly, one requires the fact that $\lambda(l_1^{(n)}) = 2^{-m}(2m+1)\binom{2m}{m}$, with $m = \left[\frac{n-1}{2}\right]$. This was established by Grünbaum in [16]. See also Daugavet's work [6]. Thirdly, one requires a calculation to show that $\|P\|$ equals this very same number; this was done by Rademacher in [43]. The proof is completed by an appeal to Theorem 9 above.

Why do we not give as examples the minimal projections of $C[a,b]$ onto Π_n or onto subspaces of spline functions? The answer is that these are not known, in general. The case of Π_1 is trivial and is subsumed by the example of piecewise-linear functions described above. However, for $n \geq 2$, the minimal projections and the projection constants of Π_n are not known. The spaces Π_n are of such importance that there would be much interest in discovering their projection constants. The search would undoubtedly lead to some new projections of great practical value, especially if they utilize simple linear functionals. The best currently available estimates of $\lambda(\Pi_n)$ are given in the next theorem.

THEOREM 13: $\lambda(\Pi_0) = \lambda(\Pi_1) = 1$, $\lambda(\Pi_2) \leq 1.2235$, and

$$\tfrac{1}{2}(\rho_n + 1/(2n+1)) \leq \lambda(\Pi_n) < \rho_n,$$

where ρ_n is the classical Lebesgue constant.

Proof: The first two equations have already been explained. For Π_2, the best-known projection is one published by Paszkowski in [40]. It is a convex combination of several interpolating projections. The general estimates for arbitrary n are obtained as follows. First, we note that the even functions in $C_{2\pi}$ form a subspace $C_{2\pi}^e$. This subspace is isometric to the space $C[-1,1]$ via the map $H: C[-1,1] \to C_{2\pi}^e$ defined by $Hx = x \circ c$, where $c(t) = \cos t$. Second, we observe that the Fourier projection F_n preserves even functions. Third, H carries Π_n into the space $°\Pi_n^e$ of nth degree cosine polynomials. A natural candidate for a projection of $C[-1,1]$ onto Π_n is therefore $H^{-1}F_n H$, and the norm of this projection is ρ_n because the norm of F_n is not decreased by restricting F_n to even functions.

For the lower bound, one proves first that if P is any projection of $C_{2\pi}^e$ onto $°\Pi_n^e$, then

$$F_0 + F_n = \frac{1}{2\pi}\int_0^{2\pi} T_\lambda P(T_\lambda + T_{-\lambda})\,d\lambda.$$

This is the analogue of the Marcinkiewicz-Berman Identity. From it we obtain $\|F_0 + F_n\| \leq 2\|P\|$. An easy calculation with the particular function $x(t) = \operatorname{sgn} D_n(t)$ yields

$$\|F_0 + F_n\| \geq [(F_0 + F_n)x](0) = 1/(2n+1) + \rho_n.$$

An exact determination of $\|F_0 + F_n\|$ can be found in Paszkowski's article [40].

Finally, we note that $H^{-1}F_n H$ is the Fourier-Tchebycheff projection, whose effect on an $x \in C[-1,1]$ is to expand x into a formal series of Tchebycheff [Čebyšev] polynomials and truncate the series. The literature on Tchebycheff polynomials is vast, and we cite only Rivlin's monograph [44] and the textbook of Fox and Parker [10]. The projection $H^{-1}F_n H$ is *not* a minimal projection of $C[-1,1]$ onto Π_n; it fails to satisfy the necessary condition of the next theorem.

THEOREM 14 (cf. [37]): *If P is a minimal projection of $C(T)$ onto a finite dimensional subspace Y and if $\|P\| > 1$, then the set of point functionals*

$$\{\hat{t} : t \in T \text{ and } \|\hat{t} \circ P\| = \|P\|\}$$

is linearly dependent over Y.

An outline of the proof of the theorem is as follows. Let K be the set of functionals which is to be proved to be dependent over Y. Let $n = \dim(Y)$, and assume that K is independent over Y. Since $\dim(Y^*) = n$, the set K can contain at most n elements. Therefore, let $\{\hat{t}_1, \ldots, \hat{t}_n\}$ be independent over Y and contain K. Select $y_1, \ldots, y_n$ in Y so that $y_i(t_j) = \delta_{ij}$. Then $Q = \sum_{i=1}^{n} \hat{t}_i \otimes y_i$ is a projection of $C(T)$ onto Y with the property that $\Lambda_Q(t) < \Lambda_P(t)$ for all $t \in K$, where $\Lambda_Q(t) = \|\hat{t} \circ Q\|$ and $\Lambda_P(t) = \|\hat{t} \circ P\|$. By a continuity and compactness argument, there exists a θ for which $\|\theta Q + (1-\theta)P\| < \|P\|$.

A special case of this theorem is of particular interest. We say that an n-dimensional subspace Y in $C(T)$ is a *Haar* subspace if 0 is the only member of Y which has n roots. Equivalently, each set of n point functionals is linearly independent over Y. It is known [37] that a Haar subspace containing constants and having dimension at least 3 in $C[a,b]$ must have projection constant greater than 1. In this case the above theorem asserts that if P is minimal, then there must exist at least $n+1$ points t having the property

$\|\hat{t} \circ P\| = \|P\|$. These special points are exactly the maximum points of the Lebesgue function of P: $\Lambda_P(t) = \|\hat{t} \circ P\|$.

Some of the results cited above enable us to answer in a simple manner the following question: Do there exist projections $P_n: C[-1,1] \to \Pi_n$ such that $\|P_n x - x\| \to 0$ for all $x \in C[-1,1]$? If such projections exist, then by the uniform boundedness principle, $\sup_n \|P_n\| < \infty$. But we have just seen that $\|P_n\| > \frac{1}{2}\rho_n > (2/\pi^2)\log n$, and thus the question must be answered negatively. Of course, the non-linear proximity maps $M_n: C[-1,1] \to \Pi_n$ do indeed have the property $\|M_n x - x\| \to 0$, and they are idempotent. Thus it must be the property of *linearity* required of P_n which spoils their convergence properties. More precisely, it must be the *additivity* requirement, for the maps M_n are homogeneous.

If P_n is a projection of $C[-1,1]$ onto Π_n, for what functions x is it true that $\|x - P_n x\| \to 0$? By Theorem 5, each x satisfying $\|P_n\| \text{dist}(x, \Pi_n) \to 0$ has the property $\|x - P_n x\| \to 0$. For an example, let P_n denote the Fourier-Tchebycheff projection. If $x \in C[-1,1]$ and if x possesses a continuous derivative, then $\text{dist}(x, \Pi_n) < k/n$ by Theorem 2. Thus

$$\|P_n\|\text{dist}(x, \Pi_n) \leq (4\pi^{-2}\log n + 1)kn^{-1}$$

and $\|x - P_n x\| \to 0$. Actually, the same conclusion can be drawn if x satisfies the weaker "Dini-Lipschitz condition": $\lim_{\delta \to 0} \omega(x;\delta) \times \log \delta = 0$.

Similar approximation-theoretic conclusions can be drawn in other situations. Consider, for example, the subspace Y of piecewise linear continuous functions with "joints" at points $a = t_1 < \cdots < t_n = b$. The interpolation projection P considered above has the property

$$\|x - Px\| \leq \omega(x;\delta), \quad \text{where } \delta = \max_{2 \leq i \leq n}(t_i - t_{i-1})$$

as is easily proved. Thus if we have a sequence of such subspaces Y_n with accompanying projections P_n and mesh sizes $\delta_n \to 0$, we can be sure that $P_n x \to x$ uniformly. Obviously then $\int_a^b (P_n x)(t)\,dt \to \int_a^b x(t)\,dt$, and this is a statement about numerical quadrature, since $\int_a^b P_n x$ is easily computed.

The projections P_n just discussed have another desirable property, namely, *monotonicity*. If $x \geqslant 0$ (i.e., $x(t) \geqslant 0$ for all t in the domain of x) then it follows that $P_n x \geqslant 0$. In some vague sense, approximation processes which preserve certain properties of a function should be more useful than those that do not. It is natural to ask then whether there exist monotone projections of $C[a,b]$ onto the *polynomial* subspaces Π_n. The answer is negative by the beautiful theorem of Bohman-Korovkin [4], [30], [31]:

THEOREM 15: *Let $\{L_n\}$ be a sequence of monotone linear operators from $C[a,b]$ into $C[a,b]$. Assume that $\|x - L_n x\| \to 0$ whenever $x \in \Pi_2$. Then $\|x - L_n x\| \to 0$ for all $x \in C[a,b]$.*

Since spline functions are beginning to usurp the role of polynomials in much of numerical analysis, questions about minimal projections onto subspaces of splines are quite important.

Let us fix our attention upon a single type of spline function. A function x defined on the real line $\mathbb{R}$ is called a *spline function of degree k* if there exist finitely many points (called "knots") $t_1 < t_2 < \cdots < t_n$ such that: (i) x is of class C^{k-1} on all of $\mathbb{R}$, (ii) in each of the intervals $(-\infty, t_1)$, $(t_1, t_2), \ldots, (t_n, \infty)$ the function x is a polynomial of degree at most k. Thus a spline function is a piecewise polynomial whose pieces have been joined together with as much smoothness as possible short of forcing the function to be globally a single polynomial. The piecewise linear continuous functions, which have occurred above as examples, are thus splines of degree 1.

If the set of knots $\{t_1, \ldots, t_n\}$ is prescribed, then the set of splines of degree k having these knots is a linear subspace of dimension $n + k + 1$ in $C(\mathbb{R})$. This is as one would expect by counting all the available coefficients, $(n+1)(k+1)$, and subtracting the number of imposed continuity conditions, nk.

If one wishes to reach high precision in approximating arbitrary continuous functions by splines, it is generally better to hold k fixed and allow n to become infinite. A popular choice is $k=3$; since the resulting cubic spline curves are of class C^2, their non-analyticity is not evident to the eye.

The class of splines described above is often reduced in dimension to n by imposing additional conditions. One way in which this occurs naturally is in the periodic case. Suppose that $a = t_0 < t_1 < \cdots < t_n = b$. Consider the cubic spline functions which have those knots and which are periodic: $x^{(i)}(a) = x^{(i)}(b)$ for $i = 0, 1, 2$. This is a linear subspace $S = S(t_0, \ldots, t_n)$ of dimension n in the space $°C[a,b]$ of periodic continuous functions.

The space S is a very useful one for purposes of approximation, and is in many ways superior to its cousin Π_{n-1}. For example, its projection constant is bounded by a number which is independent of n and independent of the spacing of knots [9]. Further, if $\delta = \max_{1 \leq i \leq n}(t_i - t_{i-1})$, then $\text{dist}(x, S) \leq c\omega(x; \delta)$ for all $x \in °C[a,b]$. Thus if we set up a sequence of such spaces S_n with accompanying δ_n converging to 0, then there exist projections P_n: $°C[a,b] \to S_n$ for which $\|x - P_n x\| \to 0$. This is in sharp contrast to the situation which prevails for the spaces Π_n.

If the distribution of knots is not too irregular, the projections P_n mentioned above can be taken to be the operation I_n of interpolation at the knots. For example, if the knots are equally spaced, then $\|I_n\| \leq 1.548$. Again, this is in marked contrast to the case of polynomial interpolation. How irregular can the spacing be and still retain the boundedness of $\|I_n\|$? This is not fully understood, but

$$\|I_n\| \geq -1 + \frac{\sqrt{3}}{36} \max_{1 \leq i \leq n} \max \left\{ \frac{t_i^{(n)} - t_{i-1}^{(n)}}{t_{i+1}^{(n)} - t_i^{(n)}}, \frac{t_{i+1}^{(n)} - t_i^{(n)}}{t_i^{(n)} - t_{i-1}^{(n)}} \right\}.$$

The right side of this inequality can become infinite with n for some choices of nodes.

5. PROJECTIONS HAVING FINITE CARRIER

From the viewpoint of numerical analysis or computer science, the only operation that can be applied directly to a function is that of evaluating it at a point. Other operations or functionals, such as integration, cannot be applied directly but must first be represented approximately by combinations of point functionals. Thus there are sound pragmatic reasons for regarding point functionals as the "building blocks" for all operations with functions.

In the space $X = C(T)$ the point functional $\hat{t}$ is defined by $\hat{t}(x) = x(t)$ for $x \in X$ and $t \in T$. The functionals $\pm \hat{t}$ are precisely the extreme points of the unit ball in X^*. By the Krein-Milman [Kreĭn-Mil'man] Theorem, every linear functional on X is a limit in the weak* topology of linear combinations of point functionals. This gives concrete meaning to the preceding informal remarks.

The *carrier* of a linear operator L on $X = C(T)$ is defined to be the smallest closed set F in T such that $Lx = 0$ whenever $x \in X$ and $x|F = 0$. (The notation $x|F$ denotes the *restriction* of x to the set F.) If L has a finite carrier, then it can be represented with finitely many point functionals: $L = \sum_{i=1}^{m} \hat{t}_i \otimes x_i$. The Krein-Milman Theorem implies that every functional is a limit of functionals with finite carrier. Is the same true of projections? Yes, and the next theorem describes the situation in detail.

THEOREM 16 (cf. [37]): *Let P be a projection of the space $X = C(T)$ onto a finite dimensional subspace Y. Then there exists a net of projections P_α from X onto Y such that*:

(1) *each P_α has finite carrier,*
(2) $\lim_\alpha \|P_\alpha x - Px\| = 0$ *for each* $x \in X$,
(3) $\lim_\alpha \|P_\alpha\| = \|P\|$,
(4) $\lim_\alpha \|\Lambda_{P_\alpha} - \Lambda_P\| = 0$.

If T is metrizable, the net can be replaced by a sequence.

(Property (4) is not stated in reference [37]; it was pointed out to me by C. Franchetti.)

The importance of the Lebesgue function of an operator L stems from the fact that $\|L\| = \|\Lambda_L\|_\infty$. This equation provides a means for computing $\|L\|$. For example, suppose that L has finite carrier: $L = \sum_{i=1}^m \hat{t}_i \otimes x_i$. An easy calculation shows that $\Lambda_L = \Sigma |x_i|$, whence $\|L\| = \max_t \Sigma |x_i(t)|$. If L is an integral operator, say $(Lx)(t) = \int_a^b k(t,s) x(s) \, ds$, then $\Lambda_L(t) = \int_a^b |k(t,s)| \, ds$.

The simplest projections with finite carrier are the interpolating projections mentioned in §2. This class is also the most thoroughly investigated one. To illustrate with a concrete problem, let us consider an interpolating projection P from $C[a,b]$ onto Π_{n-1}. The tensor form of such an operator is $P = \sum_{i=1}^{n} \hat{t}_i \otimes y_i$, with $y_i \in \Pi_{n-1}$, $a \leqslant t_1 < \cdots < t_n \leqslant b$, and $y_i(t_j) = \delta_{ij}$. The Lebesgue function $\Lambda = \sum |y_i|$ is a piecewise polynomial and has the properties $\Lambda \geqslant 1$, $\Lambda(t_i) = 1$. Hence its graph looks roughly like this:

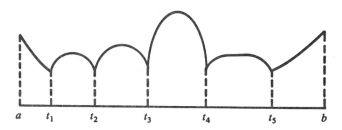

By means of an affine map $t \mapsto t' = \alpha t + \beta$, the nodes can be spread out so that $a = t'_1 < \cdots < t'_n = b$. It is not difficult to see that the interpolating projection corresponding to the new nodes has norm no greater than that of the original projection. Thus if our goal is to discover the distribution of nodes for which $\|P\|$ is least, we can assume that the endpoints of the interval are nodes. This reduces the set of parameters to $\{t_2, \ldots, t_{n-1}\}$.

It was conjectured by S. Bernstein [Bernšteĭn] in 1932 that when the minimum of $\|P\|$ is reached, the local maxima of the Lebesgue functions must be equal. (In stating this conjecture it is assumed that the endpoints are nodes; thus there are $n-1$ local maxima.) The correctness of this conjecture was finally established in July 1976 by T. A. Kilgore [27].

To some extent, the theory of finite dimensional interpolating projections can be carried over to more general subspaces $Y \subset C(T)$. But if an interpolating projection $P = \sum_{i=1}^{n} \hat{t}_i \otimes y_i$ onto Y is to exist for *any* choice of n nodes $t_1, \ldots, t_n$, then Y must have the "Haar" property previously defined. This property was introduced by Haar [17] in 1918. By a theorem of Mairhuber [35], extended by Curtis [5] and Sieklucki [47], the space $C(T)$ can contain a Haar subspace of dimension 2 or greater only if T is homeomorphic to a subset of a circular arc. Thus the hypothesis that T is a real interval does not impose a severe restriction in these problems.

The following theorem is as far as the theory of interpolating projections has progressed in this direction.

THEOREM 17 (cf. [28]): *Let Y be an n-dimensional Haar subspace containing the constant functions in $C[a,b]$. Then there exists an interpolating projection $P=\sum_{i=1}^{n}\hat{t}_i \otimes y_i$ of $C[a,b]$ onto Y, whose Lebesgue function reaches its maximum value $\|P\|$ in each of the $n+1$ intervals $(a,t_1), (t_1,t_2),\ldots,(t_n,b)$.*

A numerical procedure devised by Hayes and Powell [18] has been used to determine the nodes (in polynomial interpolation) for which the local maxima of the Lebesgue function are equal. It is not known whether this property implies that the operator have minimum norm.

Turning now to other projections with finite carrier, we outline a minimization problem that is amenable to numerical solution. Namely, one fixes the n-dimensional subspace Y in $C(T)$ and a system of points $\{t_1,\ldots,t_m\}$ in T. Consider next the family $\mathcal{C}$ of all projections P of $C(T)$ onto Y which can be expressed using only the prescribed point functionals $\hat{t}_i$. This family $\mathcal{C}$ is a linear manifold (translate of a linear subspace) in the space $\mathcal{L}(X,Y)$. Its dimension is not greater than $n(m-n)$, and this number *is* the dimension if Y is a Haar subspace. If $m=n$, the set $\mathcal{C}$ contains just one element, namely, the interpolation projection having nodes $t_1,\ldots,t_n$. If $m>n$, then one can ask for a determination of the minimal elements in $\mathcal{C}$.

This minimization problem can be put into the form of a linear programming problem or into a problem of Tchebycheff approximation. Since the latter formulation is more interesting, as well as more favorable for numerical solution, an outline of it will be given.

The problem can be treated in a more general setting as follows. We fix Banach spaces $Y \subset X$, $V \subset X^*$ with $\dim Y = n < \infty$, $\dim V = m < \infty$, and V total over Y. We seek a minimal projection $P:X \to Y$ among all those for which $P^*X^* \subset V$.

Select a basis $\{f_1,\ldots,f_m\}$ for V such that $\{f_1,\ldots,f_n\}$ is total over Y. Then there is a projection of the form $Q = \sum_{i=1}^{n} f_i \otimes y_i$ from X onto Y. Any other projection in the class being considered must be of the form $P = Q - L$ where $LX \subset Y$, $LY = 0$, and $L^*X^* \subset V$.

Hence $L = \sum_{i=1}^{m} f_i \otimes u_i$ for appropriate $u_i \in Y$. The condition $LY = 0$ is equivalent to

$$0 = Ly_j = \sum_{\nu=1}^{m} f_\nu(y_j) u_\nu = \sum_{\nu=1}^{m} f_\nu(y_j) \sum_{i=1}^{n} f_i(u_\nu) y_i \quad (1 \leq j \leq n).$$

Since the set $\{y_1, \ldots, y_n\}$ is linearly independent, then we must have $\sum_{\nu=1}^{n} f_i(u_\nu) f_\nu(y_j) = 0$ for $1 \leq i, j \leq n$. This condition can be written in matrix form as $BA = 0$, with $B_{i\nu} = (f_i(u_\nu))$ and $A_{\nu j} = (f_\nu(y_j))$. Thus an admissible set $\{u_1, \ldots, u_m\}$ can be obtained from any matrix B satisfying $BA = 0$ via the equations $u_\nu = \sum_{i=1}^{n} B_{i\nu} y_i$ ($1 \leq \nu \leq m$). Now the rows of B must be orthogonal to the columns of A, and the orthogonal complement of the columns space of A is generated by the rows of the matrix

$$D_{ij} = -f_{n+i}(y_j) \quad \text{for } 1 \leq j \leq n,$$
$$= \delta_{j, n+i} \quad \text{for } n < j \leq m.$$

Thus each row of B must be a linear combination of rows of D, and the matrix equation $B = CD$ is valid with C an unconstrained $n \times (m-n)$-matrix. Thus

$$P = Q - L = \sum_{i=1}^{m} f_i \otimes \left[y_i - \sum_{\nu=1}^{n} \sum_{j=1}^{m-n} C_{\nu j} D_{ji} y_\nu \right]$$

in which we have set $y_{n+1} = \cdots = y_m = 0$.

Now define the compact set

$$Q = \{(x, g) : x \in \text{ext} S(X^*), \ g \in \text{ext} S(Y^*)\}$$

where ext denotes the set of extreme points and $S(X)$ is the unit sphere of a space X. The norm of P can be expressed as

$$\|P\| = \sup_{(x,g) \in Q} |g(Px)|$$

$$= \sup_{(x,g) \in Q} \left| \sum_{i=1}^{m} f_i(x) \left[g(y_i) - \sum_{\nu=1}^{n} \sum_{j=1}^{m-n} C_{\nu j} D_{ji} g(y_\nu) \right] \right|$$

$$= \sup_{(x,g) \in Q} \left| F(x, g) - \sum_{\nu=1}^{n} \sum_{j=1}^{m-n} C_{\nu j} H_{\nu j}(x, g) \right|$$

with

$$F(x,y) = \sum_{i=1}^{m} f_i(x)g(y_i) \quad \text{and} \quad H_{\nu j}(x,g) = \sum_{i=1}^{m} D_{ji}f_i(x)g(y_\nu).$$

Thus the minimization of $\|P\|$ is equivalent to finding a best approximation (in a supremum norm) of a function F by a linear combination of functions $H_{\nu g}$. There exist good algorithms for such problems, for example those of Bartels and Golub [2].

We conclude this section by citing a theorem from [37] which characterizes the minimal projection with carrier contained in a prescribed finite set. Let Y be a finite dimensional subspace of $X \equiv C(T)$, and let $F = \{t_1, \ldots, t_m\}$ be a prescribed finite subset of T. Let $\mathcal{B}(F)$ be the family of all projection from X onto Y whose carriers are contained in F.

THEOREM 18: *In order that element P be minimal in $\mathcal{B}(F)$, it is necessary and sufficient that there exist $f_1, \ldots, f_m \in X^*$ such that*
  (i) $\mathrm{carr}(f_i) \subset \{t : \|\hat{t} \circ P\| = \|P\|\}$
  (ii) $\||f_1| \vee \cdots \vee |f_m|\| = 1$
  (iii) $\sum_{i=1}^{m} f_i(y_i) = \|P\|$ whenever $\sum_{i=1}^{m} \hat{t}_i \otimes y_i \in \mathcal{B}(F)$.

6. ORTHOGONAL PROJECTIONS IN $C(T)$

Example 2 in §2 described a family of orthogonal projections from $C[a,b]$ onto Π_{n-1}. Here we shall extend this concept somewhat and discuss the accompanying extremum problems. The material is taken from [11].

Consider a subspace Y in $X = C(T)$ and a projection P of X onto Y. Under what conditions should we term P an "orthogonal projection"? In the classical theory there should exist an inner product on X such that $(x - Px) \perp Y$ for all x; that is, $\langle x - Px, y \rangle = 0$ for all $x \in X$ and for all $y \in Y$. In order to have an existence theorem for minimal orthogonal projections, the definition should be generalized as follows. The projection P will be termed *orthogonal* if there exists a positive functional f of norm 1 in X^* such that $f((x - Px)y) = 0$ for all $x \in X$ and for all $y \in Y$. Of course, xy

denotes the function $t \mapsto x(t)y(t)$. Now it is not difficult to apply the existence theorem of §3 and conclude that among all the orthogonal projections from $C(T)$ onto a finite dimensional subspace Y, there is one of minimal norm. The class of orthogonal projections thus defined is quite large, and contains the interpolating projections. (In the definition, let f be a point-functional corresponding to a node.) The reason that the class is so large is that we have not imposed the condition ordinarily met in the classical case. Namely, we have not insisted that the equation $\langle x,y \rangle = f(xy)$ define a true inner product in X. The missing condition is that $f(x^2) > 0$ for each nonzero function x in X. Unfortunately, when this condition *is* imposed, the resulting subset of orthogonal projections is no longer closed in the w*o-topology, and the existence of a minimal element is no longer assured.

An interesting subclass of orthogonal projections which is w*o-closed can be obtained as follows. Fix $f \in X^*$ which is such that $f \geqslant 0$, $\|f\| = 1$, and $f(y^2) > 0$ for all $y \in Y \setminus 0$. By the Gram-Schmidt process one can construct a basis for Y which is f-orthonormal: $f(y_i y_j) = \delta_{ij}$ ($1 \leqslant i,j \leqslant n$). Then an orthogonal projection is obtained by defining $P_f x = \sum_{i=1}^n f(xy_i) y_i$. Now consider the family $\mathcal{Q}_f$ of all projections P_g, with $g \in X^*$, $g \geqslant 0$, $\|g\| = 1$, and $g(y_i y_j) = \delta_{ij}$. The family $\mathcal{Q}_f$ is sometimes a singleton, but usually it is infinite dimensional. It is w*o-closed, and so contains a minimal element.

The Lebesgue function of the projection P_f is $\Lambda(t) = f(|k_t|)$ with $k_t = \sum_{i=1}^n y_i(t) y_i$, the "kernel" of P_f. Observe that every projection in $\mathcal{Q}_f$ has the same kernel. The kernel is independent of the basis, and has the property $\hat{t} P_g x = g(xk_t)$ for all $P_g \in \mathcal{Q}_f$, $t \in T$, $x \in X$. In [11], the following characterization of minimal elements in $\mathcal{Q}_f$ is proved:

THEOREM 19: *In order that P_f be minimal in $\mathcal{Q}_f$, it is necessary and sufficient that no element h in X^* have the three properties $h \leqslant f$, $h(y_i y_j) = 0$ for $1 \leqslant i,j \leqslant n$, and $h(|k_t|) > 0$ for all t satisfying $f(|k_t|) = \|P_f\|$.*

The proof is not difficult. For example, the sufficiency is proved as follows. If P_f is not minimal, then there exists $g = f - h \in X^*$ such that $P_g \in \mathcal{Q}_f$ and $\|P_g\| < \|P_f\|$. Then $h \leqslant f$ because $g \geqslant 0$. Next

$h(y_i y_j) = 0$ since $g(y_i y_j) = \delta_{ij} = f(y_i y_j)$. Finally, if $f(|k_t|) = \|P_f\|$ then

$$g(|k_t|) = \Lambda_{P_g}(t) \leq \|P_g\| < \|P_f\| = f(|k_t|).$$

One corollary of this result concerns projections from $C[a,b]$ onto Π_n. Assume that $f(x^2) > 0$ for all $x \in C[a,b] \setminus 0$. If P_f is minimal in $\mathcal{Q}_f$ then its Lebesgue function must attain its maximum value at infinitely many points. On the other hand, if the Lebesgue function is constant, then P_f is minimal in $\mathcal{Q}_f$.

As an illustration, consider the Fourier-Tchebycheff projection, described in §4. This is an orthogonal projection P_f when the functional f is defined to be $f(x) = \int_{-\pi}^{\pi} x(\cos\theta) d\theta$, $x \in C[-1,1]$. Although this is a very useful projection and is close to being minimal, it is not even minimal in its class $\mathcal{Q}_f$. This can be verified by using the previous corollary. The Lebesgue function of this projection reaches its maximum at just two points of the interval $[-1,1]$.

7. PROJECTIONS IN $\mathcal{L}_1$-SPACES

Attention is turned here to the problem of determining minimal projections from a space $X = \mathcal{L}_1(T, \Sigma, \mu)$ onto a given finite dimensional subspace Y. Even the elementary case of $\mathcal{L}_1[0,1]$ and $Y = \Pi_1$ presents some surprises. The results discussed here are taken from [12].

The space $\mathcal{L}_1(T, \Sigma, \mu)$ consists of all integrable functions on the measure space (T, Σ, μ), with norm defined by $\|x\|_1 = \int_T |x(t)| d\mu(t)$. Of course, functions equal to each other almost everywhere are deemed to be identical. It is assumed that $X^* = \mathcal{L}_\infty(T, \Sigma, \mu)$, which is always true if (T, Σ, μ) is σ-finite. The space $\mathcal{L}_\infty(T, \Sigma, \mu)$ consists of all essentially bounded measurable functions on T with norm $\|u\|_\infty = \operatorname{ess\,sup}|u(t)|$. The equation of duality is

$$u(x) = (u, x) = \int_T u(t) x(t) d\mu(t) \qquad (x \in \mathcal{L}_1, u \in \mathcal{L}_\infty).$$

If Y is a finite dimensional subspace of X and if $\{y_1, \ldots, y_n\}$ is a basis for Y, then each projection of X onto Y is of the form

$$P = \sum_{i=1}^n u_i \otimes y_i \quad \text{i.e.,} \quad Px = \sum_{i=1}^n (x, u_i) y_i$$

for appropriate $u_i \in \mathcal{L}_\infty$ satisfying $(u_i, y_j) = \delta_{ij}$.

In order to discuss the minimization of $\|P\|$, a convenient formula for this norm is needed. It turns out to be

$$\|P\| = \|\Lambda_P\|_\infty, \quad \text{where} \quad \Lambda_P(t) = \|\Sigma u_i(t) y_i\|_1.$$

The function Λ_P is termed the *Lebesgue function* of P just as in the case of $C(T)$. It depends only on P, not on the particular representation of P. Observe that $\Lambda_P \in \mathcal{L}_\infty$ in this setting.

Now consider a finite dimensional smooth subspace Y in X. Smoothness is interpreted in the Banach-space sense: at each boundary point of the intersection of Y with the unit ball of X there is only one supporting hyperplane of the unit ball. In measure theoretic terms, it means that each nonzero element of Y is almost everywhere different from 0. A crucial theorem is then as follows.

THEOREM 20: *Let (T, Σ, μ) be a nonatomic measure space for which $\mathcal{L}_1^* = \mathcal{L}_\infty$. Let Y be a smooth and finite dimensional subspace in $\mathcal{L}_1$. If P is a minimal projection of $\mathcal{L}_1$ onto Y, then the Lebesgue function of P is constant.*

In outline the proof proceeds as follows. If Λ_P is not constant, then there is a number δ such that the sets $A = \{t : \Lambda_P(t) \geq \|P\| - \delta\}$ and $B = \{t : \Lambda_P(t) < \|P\| - \delta\}$ both have positive measure. It can be shown, by an inductive proof on the dimension of Y, that there exist disjoint measurable sets $B_1, \ldots, B_n$ in B and a basis $y_1, \ldots, y_n$ for Y such that $\int_{B_i} y_j = \delta_{ij}$. Select $u_1, \ldots, u_n \in \mathcal{L}_\infty$ so that $P = \sum_{i=1}^n u_i \otimes y_i$. Define $C = T \setminus (B_1 \cup \cdots \cup B_n)$, and consider a perturbation of the u_i:

$$\tilde{u}_i(t) = u_i(t) + \epsilon_{ij} \quad \text{on } B_j,$$
$$= (1 - \epsilon) u_i(t) \quad \text{on } C.$$

If $\epsilon_{ij} = \epsilon \int_C u_i y_j$, then the operator $\tilde{P} = \Sigma \tilde{u}_i \otimes y_i$ will be a projection of $\mathcal{L}_1$ onto Y. If ϵ is sufficiently small, then $\|\tilde{P}\| < \|P\|$.

As a corollary, one can prove that the minimal projection onto Y is unique, if in addition to the hypotheses of the theorem one assumes that Y is rotund (strictly convex). Indeed, if $P = \Sigma u_i \otimes y_i$ and $Q = \Sigma v_i \otimes y_i$ are minimal projections, then so is $\frac{1}{2}(P+Q)$. Its Lebesgue function must be constant, and so for almost all t, we have

$$\|\tfrac{1}{2}\Sigma u_i(t)y_i + \tfrac{1}{2}\Sigma v_i(t)y_i\|_1 = \|\Sigma u_i(t)y_i\|_1 = \|\Sigma v_i(t)y_i\|_1.$$

By the strict convexity of Y, we have $\Sigma u_i(t)y_i = \Sigma v_i(t)y_i$ and so $u_i = v_i$.

Another necessary condition that must be fulfilled by a minimal projection $P = \Sigma_1^n u_i \otimes y_i$ is:

(C) There do not exist $v_1, \ldots, v_n \in \mathcal{L}_\infty \cap Y^\perp$ such that

$$\operatorname*{ess\,sup}_t \int \sum_{i=1}^n v_i(t)y_i(s)\operatorname{sgn}\left[\sum_{j=1}^n u_j(t)y_j(s)\right] ds < 0.$$

This can be proved under the supposition that (T, Σ, μ) is a finite and nonatomic measure space, and that Y is finite dimensional and smooth. The proof is rather technical and will not be given here. The point of introducing the condition (C) is that the two necessary conditions which have been discussed are together sufficient for the minimality of P.

If the characterization theorem just outlined is applied to the polynomials of degree 1 in $\mathcal{L}_1[-1, 1]$ with Lebesgue measure, then (after somewhat lengthy calculations) one arrives at the unique minimal projection in the following form:

$P = \Sigma_{i=1}^2 u_i \otimes y_i$ with $y_1(t) = 1$, $y_2(t) = t$, $u_1(t) = \alpha\beta(\beta^2 + t^2)^{-1/2}$, and $u_2(t) = \alpha(1 + |t|(\beta^2 + t^2)^{-1/2})\operatorname{sgn} t$.

In these equations α and β are constants determined as follows. First solve the equation $2\xi(1 + \xi - \xi^2)\log\xi = \xi^2 - 1$ for the root ξ in the interval $0 < \xi < 1$. Then define $\beta = 2\xi(1-\xi^2)^{-1}$ and $\alpha = -(2\beta \log \xi)^{-1}$. The norm of the projection is $2\alpha = 1.2204049$.

8. SOME OPEN PROBLEMS

1. What are the projection constants of the polynomial subspaces Π_n under the supremum norm in $[a,b]$?
2. What are the projection constants of the common spline spaces under the supremum norm? In particular, what is the projection constant of the periodic cubic splines on an interval with equidistant knots?
3. In trigonometric interpolation in the space $C_{2\pi}$, is the norm of the interpolation operator a minimum when the interpolation nodes are equidistant?
4. What n-dimensional Banach space has the greatest projection constant?
5. Consider the cubic spline interpolation operator described in §4. Is its norm minimized by taking equidistant knots?
6. What condition on the spacing of knots is necessary and sufficient for the uniform boundedness of the cubic spline interpolants described in §4?
7. Find an elegant and useful characterization of the minimal projections from $C(T)$ onto a prescribed subspace Y.

Acknowledgments. When there were matters that I did not understand, I turned for help to my colleagues Elton Lacey and Carlo Franchetti. I thank them for their assistance and their patience. They are not to be blamed for my errors.

REFERENCES

1. B. Bajšanski and R. Bojanić, "A note on approximation by Bernstein polynomials," *Bull. Amer. Math. Soc.*, **70** (1964), 675–677.
2. R. H. Bartels and G. H. Golub, "Stable numerical methods for obtaining the Chebyshev solution to an overdetermined system of linear equations," *Comm. ACM*, **11** (1968), 401–406, 428–430; ibid., **12** (1969), 326.
3. D. L. Berman, "On the impossibility of constructing a linear polynomial operator furnishing an approximation of the order of best approximation," *Dokl. Akad. Nauk SSSR*, **120** (1958), 143–148 (Russian).

4. H. Bohman, "On approximation of continuous and of analytic functions," *Ark. Mat.*, 2 (1952), 43–56.
5. P. C. Curtis, "N-parameter families and best approximation," *Pacific J. Math.*, 9 (1959), 1013–1027.
6. I. K. Daugavet, "Some applications of the generalized Marcinkiewicz-Berman identity," *Vestnik Leningrad. Univ.*, 23 (1968), no. 19, 59–64 (Russian) = *Vestnik Leningrad. Univ. Math.*, 1 (1974), 321–327; MR **40** # 626.
7. W. J. Davis, "Remarks on finite-rank projections," *J. Approximation Theory*, 9 (1973), 205–211.
8. M. M. Day, *Normed Linear Spaces*, Springer-Verlag, Berlin, 1962.
9. C. de Boor, "On uniform approximation by splines," *J. Approximation Theory*, 1 (1968), 219–235.
10. L. Fox and I. B. Parker, *Chebyshev Polynomials in Numerical Analysis*, Oxford University Press, 1968.
11. C. Franchetti and E. W. Cheney, "Orthogonal projections in spaces of continuous functions," *CNA*, 113, Center for Numerical Analysis, Univ. of Texas at Austin, Sept. 1976; *J. Math. Anal. Appl.*, 63 (1978), 253–264.
12. _____, "Minimal projections in $\mathcal{L}_1$-spaces," *Duke Math. J.*, 43 (1976), 501–510.
13. P. V. Galkin, "Estimates of Lebesgue constants," *Trudy Mat. Inst. Steklov.*, 109 (1971), 3–5.
14. D. J. H. Garling and Y. Gordon, "Relations between some constants associated with finite dimensional Banach spaces," *Israel J. Math.*, 9 (1971), 346–361.
15. M. Golomb, "Optimal and nearly-optimal linear approximation," in *Approximation of Functions*, H. L. Garabedian, ed., Elsevier, New York, 1965, pp. 83–100.
16. B. Grünbaum, "Projection constants," *Trans. Amer. Math. Soc.*, 95 (1960), 451–465.
17. A. Haar, "Die Minkowskische Geometrie und die Annäherung an stetige Funktionen," *Math. Ann.*, 78 (1918), 294–311.
18. W. E. Hayes and M. J. D. Powell, unpublished data, Atomic Research Establishment, Harwell, England, 1969.
19. R. B. Holmes, "Best approximation by normal operators," *J. Approximation Theory*, 12 (1974), 412–417; MR **50** # 8128.
20. R. Holmes, B. Scranton, and J. Ward, "Best approximation by compact operators, II," *Bull. Amer. Math. Soc.*, 80 (1974), 98–102; MR **50** # 8137.
21. Y. Ikebe, "Generalizations of the Alaoglu theorem with applications to approximation theory," *Proc. Japan Acad.*, 44 (1968), 439–445.
22. J. R. Isbell and Z. Semadeni, "Projection constants and spaces of continuous functions," *Trans. Amer. Math. Soc.*, 107 (1963), 38–48.
23. F. John, "Extremum problems with inequalities as subsidiary conditions," in *Courant Anniversary Volume*, Interscience, New York, 1948, pp. 187–204.
24. M. I. Kadec, "On Lozinski-Kharshiladze systems," *Uspehi Mat. Nauk*, 18 (1963), no. 5 (113), 167–169.
25. M. I. Kadec and B. S. Mitjagin, "Complemented subspaces in Banach spaces,"

Uspehi Mat. Nauk, **28** (1973), no. 6 (174), 77-94 = *Russian Math. Surveys*, **28** (1973), 77-95.

26. M. I. Kadec and M. G. Snobar, "Some functionals over a compact Minkowski space," *Mat. Zametki*, **10** (1971), 453-457 = *Math. Notes*, **10** (1971), 694-696; MR **45** # 861.
27. T. A. Kilgore, "A proof of Bernstein's conjecture in the theory of interpolation" (to appear).
28. T. A. Kilgore and E. W. Cheney, "A theorem on interpolation in Haar subspaces," *Aequationes Math.*, **14** (1975), 391-400.
29. A. N. Kolmogorov, "Über die beste Annäherung von Funktionen einer gegebenen Funktionenklasse," *Ann. of Math.*, (2) **37** (1936), 107-111.
30. P. P. Korovkin, *Linear Operators and Approximation Theory*, Fizmatgiz, Moscow, 1959 (Russian); translation, Hindustan Publishing Corp., Delhi, 1960.
31. _____, "On convergence of linear positive operators in the space of continuous functions," *Dokl. Akad. Nauk SSSR*, **90** (1953), 961-964 (Russian).
32. H. E. Lacey, *The Isometric Theory of Classical Banach Spaces*, Springer-Verlag, Berlin, 1974.
33. H. Lebesgue, "Sur les intégrales singulières," *Ann. Fac. Sci. Univ. Toulouse*, Ser. 3, **1** (1909), 25-117.
34. G. G. Lorentz, *Bernstein Polynomials*, Univ. of Toronto Press, Toronto, 1953.
35. J. C. Mairhuber, "On Haar's theorem concerning Chebychev approximation problems having unique solutions," *Proc. Amer. Math. Soc.*, **7** (1956), 609-615.
36. J. Marcinkiewicz, "Quelques remarques sur l'interpolation," *Acta Litt. Sci. Szeged*, **8** (1937), 127-130.
37. P. D. Morris and E. W. Cheney, "On the existence and characterization of minimal projections," *J. Reine Angew. Math.*, **270** (1974), 61-76.
38. F. J. Murray, "On complementary manifolds and projections in spaces L_p and l_p," *Trans. Amer. Math. Soc.*, **41** (1937), 138-152.
39. V. P. Odinec, "On uniqueness of minimal projections in Banach spaces," *Dokl. Akad. Nauk SSSR*, **220** (1975) = *Sov. Math. Dokl.*, **16** (1975), 151-154.
40. S. Paszkowski, *Polynômes et séries de Tchebycheff*, Séminaire d'Analyse Numérique, Univ. Grenoble, 19 May 1971.
41. R. R. Phelps, "Uniqueness of Hahn-Banach extensions and unique best approximation," *Trans. Amer. Math. Soc.*, **95** (1960), 238-255.
42. E. D. Positselskii [Posicelśkiĭ], "Projection constants of symmetric spaces," *Mat. Zametki*, **15** (1974), 719-727 (Russian) = *Math. Notes*, **17** (1974), 430-435.
43. H. Rademacher, "Einige Sätze über Reihen von allgemeinen Orthogonalfunktionen," *Math. Ann.*, **87** (1922), 112-138.
44. T. J. Rivlin, *The Chebyshev Polynomials*, Wiley, New York, 1974.
45. N. A. Sapogov, "Norms of linear polynomial operators," *Soviet Math. Dokl.*, **3** (1962), 602-604.
46. R. Scholz, "Abschätzungen linearer Durchmesser in Sobolev- und Besov-Raümen," *Manuscripta Math.*, **11** (1974), 1-14; MR **50** # 8054.
47. K. Sieklucki, "Topological properties of sets admitting the Tschebycheff systems," *Bull. Acad. Polon.*, **6** (1958), 603-606.
48. P. C. Sikkema, "Der Wert einiger Konstanten in der Theorie der Approxima-

tion mit Bernstein-Polynomen," *Numer. Math.*, **3** (1961), 107–116.
49. S. B. Stečkin, "Best approximation of linear operators," *Mat. Zametki*, **1** (1967), 91–99.
50. V. M. Tihomirov, "Widths of sets in functional spaces and the theory of best approximations," *Uspehi Mat. Nauk*, **15**, no. 3 (1960), 75–112 = *Russian Math. Surveys*, **15** (1960), 75–112.
51. _____, "Some problems in approximation theory," *Dokl. Akad. Nauk*, **160** (1965), 774–777 (Russian) = *Soviet Math.*, **6** (1965), 202–205.

COMPLEMENTABLY UNIVERSAL SEPARABLE BANACH SPACES: AN APPLICATION OF COUNTEREXAMPLES TO THE APPROXIMATION PROBLEM

William B. Johnson

1. INTRODUCTION

In [2] Per Enflo gave the first example of a Banach space which fails to have the approximation property, thus solving the best-known open problem in the geometric theory of Banach spaces. It is not surprising that Enflo's results and techniques have been used to solve many other problems in Banach spaces and such closely related fields as uniform algebras. In this paper we will describe, for example, how A. Szankowski and the author [8] applied extensions developed by Davie [1] and Figiel [3] of Enflo's result and techniques, to prove the nonexistence of a separable Banach space which is complementably universal for the class of all separable Banach spaces.

The research for this paper was supported in part by NSF-MPS 72-04634-A03.

Before stating any results more formally, we will explain the background material which motivated the approximation problem and problems on complementably universal Banach spaces. Recall that an *operator* T (which will always mean bounded linear operator) from a Banach space X into a Banach space Y is called *compact* provided that T carries the *unit ball* $B_X \equiv \{x \in X : \|x\| \leq 1\}$ of X into a subset of Y which has compact closure. If the operator $T: X \to Y$ has *finite rank*, i.e., if $\dim Tx < \infty$, then TB_X is finite dimensional and bounded, so that T is compact. Now if T is an operator which can be approximated uniformly by compact operators (i.e., if $\|T - T_n\| \equiv \sup\{\|Tx - T_n x\| : x \in B_X\} \to 0$ as $n \to \infty$, for some sequence (T_n) of compact operators), then T is compact. Hence the compact operators on X contain every operator which can be approximated uniformly by finite rank operators. Whether the converse is true was a classical problem, known as "the approximation problem."

Grothendieck [6] formulated the concept of the approximation property: A Banach space Y has the *approximation property* (a.p., for short) provided that for each compact subset K of Y and each $\epsilon > 0$, there is a finite rank operator $T: Y \to Y$ which satisfies $\sup\{\|y - Ty\| : y \in K\} < \epsilon$. Grothendieck proved that Y has the approximation property if and only if, for each Banach space X, every compact operator from X into Y can be approximated uniformly by finite rank operators from X to Y. One direction of the equivalence is very easy. Suppose Y has the a.p. and $T: X \to Y$ is compact. Let K be the compact set $\overline{TB_X}$ and let $\epsilon > 0$ be arbitrary. If S is a finite rank operator from Y to Y which satisfies

$$\sup\{\|Sk - k\| : k \in K\} < \epsilon,$$

then the finite rank operator $ST: X \to Y$ satisfies $\|ST - T\| < \epsilon$. Thus T can be uniformly approximated by finite rank operators.

In [2], Enflo constructed a Banach space which fails to have the a.p. of Grothendieck and thereby solved the classical approximation problem in the negative. Actually, Enflo's example, X, even fails to have the *compact approximation property* (c.a.p., for short). That is, there is a compact subset K of X and an $\epsilon > 0$, so that $\sup\{\|Tx - x\| : x \in K\} \geq \epsilon$ for every compact operator $T: X \to X$. In Section 4 we present Figiel's construction [3] (which is a variation of an earlier construction of Figiel and Pełczyński [5]) of subspaces of l_p ($2<p<\infty$) which fail to have the c.a.p.

We turn attention now to the (apparently unrelated) topic of universal Banach spaces. A useful classical theorem of Banach and Mazur states that the Banach space $C[0,1]$ of continuous functions from $[0,1]$ into the scalars under the supremum norm $\|f\| = \sup\{|f(t)|:t\in[0,1]\}$ is universal for the class of all separable Banach spaces; i.e., every separable Banach space is isometrically isomorphic to a (closed linear) subspace of $C[0,1]$. Since $C[0,1]$ is separable (the polynomials with rational coefficients are dense in $C[0,1]$), one can say that the class of separable Banach spaces contains a universal member.

However, there generally is not a "nice" imbedding of a separable Banach space X into $C[0,1]$. To make this precise, we need recall the concept of complemented subspace. A closed subspace Y of the Banach space X is said to be *complemented* provided that there is a projection P from X onto Y. (Recall that an operator P is called a *projection* provided $P^2 = P$.) Of course, a projection P induces a direct sum decomposition $X = Y + Z$, where the closed subspace Z (= kernel P) satisfies $Y \cap Z = \{0\}$. It is easy to see that X is then isomorphic to the direct sum $Y \oplus Z$, where $Y \oplus Z$ is normed by any common, direct sum norm—e.g., $\|(y,z)\| = (\|y\|^2 + \|z\|^2)^{1/2}$.

Evidently, a complemented subspace of X is more nicely imbedded into X than an uncomplemented subspace. For example, one can argue that Hilbert space operator theory is so satisfactory simply because every subspace is complemented. However, such nice behavior is restricted to Hilbert space, since Lindenstrauss and Tzafriri [12] proved the beautiful result that, if the Banach space X is not isomorphic (= linearly homeomorphic) to a Hilbert space, then X contains a subspace which is not complemented. This means that every Banach space other than Hilbert space has subspaces which are "badly imbedded." In particular, $C[0,1]$ has uncomplemented subspaces. Of course, one might hope that, even though the subspace X of $C[0,1]$ is not complemented, X is isometrically isomorphic, or at least *isomorphic* (that is to say, *linearly homeomorphic*) to another subspace of $C[0,1]$ which is complemented in $C[0,1]$. As it happens, very few Banach spaces are isomorphic to complemented subspaces of $C[0,1]$. Suppose X is isomorphic to a complemented subspace of $C[0,1]$. Pełczyński [15] proved that c_0 is isomorphic to a subspace of X. Since c_0 is not reflexive, neither is X; so $C[0,1]$ contains no reflexive complemented subspaces (a fact which actually was known prior to Pełczyński's work). Rosenthal [20] proved that if X^* is not separable, then X is isomorphic to $C[0,1]$. This means that $C[0,1]$ has no "big" complemented subspaces other than itself. It is a well-known conjecture that if X^* is separable, then X is isomorphic to $C(K)$ for some countable compact metric space K. Lewis and Stegall [11] showed that X^* must be isomorphic to l_1, which lends support to the conjecture, since $C(K)^*$ is isomorphic (even isometrically isomorphic) to l_1 when K is countable (cf. [18]).

The above discussion indicates, on the one hand, that there is a fruitful positive theory concerning the "nicely imbedded," or complemented, subspaces of $C[0,1]$ and also suggests a problem in the general theory of Banach spaces: *Does there exist a separable Banach space Z which is "complementably universal" for the class of separable Banach spaces; i.e., every separable Banach space is isomorphic to a complemented subspace of Z?*

In Section 3 we show how counterexamples to the c.a.p. in l_p yield an easy negative answer to this problem. In [8], a somewhat stronger result is proved: there is no separable Banach space which is complementably universal for the class of all separable Banach spaces which have the a.p. (It is perhaps interesting to note that the proof of this latter negative result also makes use of counterexamples to the approximation problem, even though it is concerned only with spaces which possess the a.p.!)

If one restricts attention to rather good spaces, then there are positive results in the direction suggested by the "complementably universal" question. A separable Banach space is said to have the *bounded approximation property* (b.a.p., for short) provided that there is a sequence (T_n) of finite rank operators on X such that $\|T_n x - x\| \to 0$ for each $x \in X$. Such a sequence (T_n) is uniformly bounded (by the uniform boundedness principle); hence if $K \subseteq X$ is compact, $\lim_{n\to\infty} \sup_{x \in K} \|T_n x - x\| = 0$; i.e., X has the a.p. However, there are spaces which have the a.p. but fail the b.a.p. (cf. [4]). In Section 2 we give a proof of the result of Kadec [9] and

Pełczyński [16], [17] that there is a separable Banach space U which has the b.a.p. (even a Schauder basis) and which is complementably universal for the class of all separable Banach spaces which have the b.a.p. Further, U is unique in the sense that if Y is any other separable Banach space with the b.a.p. which is complementably universal for the class of separable Banach spaces with the b.a.p., then Y and U are isomorphic. This space U has many other interesting properties, some of which we will investigate here. For example, U is isomorphic to $(\Sigma U)_{l_p}$ for each $1 \leq p < \infty$; U^* is isomorphic to $(\Sigma E_n)_{l_\infty}$ for some sequence (E_n) of finite dimensional spaces [7]; U^* fails the approximation property (cf. [7]); in the case of complex scalars, U is isomorphic to a closed subalgebra of C(Cantor set) (cf. [13]).

We should mention here the organization of this paper. Sections 2, 3, and 4 are independent of each other. Section 2 gives proofs of the known positive results on complementably universal separable Banach spaces discussed above, and Section 3 contains a proof of the negative result that there is no separable Banach space that is complementably universal for all separable Banach space. Section 4 contains the argument that l_p ($2 < p \leq \infty$) has a subspace which fails to have the c.a.p.

Rather than give detailed proofs of results, we have used a problem-with-hints format. The hints contain all the important conceptual ideas for working the problems, and some of the analytical calculations. The reader who has had an introductory course in functional analysis can probably follow the discussion on complementably universal Banach spaces in Sections 2 and 3 without filling in the details that are left out. However, in order to understand the construction of counterexamples to the c.a.p. in Section 4, the reader will have to work most of the problems.

2. POSITIVE RESULTS ON UNIVERSAL BANACH SPACES

We begin with an outline of the proof that every separable Banach space is isometrically isomorphic to a subspace of $C[0, 1]$. Next, we use this classical result to construct Pełczyński's universal basis space U, which has the property that every separable Banach space possessing a Schauder basis is isomorphic to a complemented subspace of U. (The proof outlined here is due to G. Schechtman [21].) Finally we give a proof (due to Pełczyński and Wojtaszczyk [19], and Pełczyński [17]) that every separable Banach space which has the b.a.p. is isomorphic to a complemented subspace of U.

At the end of this section are a few comments concerning some peculiar properties of Pełczyński's space U. It is perhaps of particular interest that Pełczyński's space U is, in the case of complex scalars, isomorphic to a subalgebra of C(Cantor set). This follows from Milne's [13] remarkable but simple result that every separable Banach space is isomorphic to a complemented subspace of a uniform algebra. (This more or less means that, despite the extra structure, the general uniform algebra behaves as badly as the general Banach space.)

We turn now to our first problem (the reader who is familiar with this classical result should skip to Problem 2.2).

PROBLEM 2.1: *Let X be a separable Banach space. Then X is isometrically isomorphic to a subspace of $C[0,1]$.*

One sequence of steps that leads to a solution of Problem 1 is given below. In preparation for this, we need to recall the concept of the weak* or w^* topology on X^*. (Recall that X^*, the dual or conjugate to X, is the space of all bounded linear functionals on X, with the norm given by $\|x^*\| = \sup\{|x^*(x)| : x \in B_X\}$.) Since X^* is a subset of F^X, the space of all scalar-valued functions with domain X, we can relativize the product topology on F^X to X^*. (Here F denotes the field of scalars.) This topology on X^* is called the w^* topology. Thus, a sequence, or, more generally, a net $\{x_\alpha^*\}$, in X^* w^*-converges to x^* in X^* provided the net of scalars $\{x_\alpha^*(x)\}$ converges to $x^*(x)$ for every $x \in X$. Finally, we wish to recall the useful Banach-Alaoglu theorem: *Given any normed space X, B_{X^*} is compact in the w^* topology.* (The reader who is not familiar with this theorem might treat it as a problem. Note that B_{X^*} is contained in the compact product space $\Pi_{x^* \in X^*}\{a \in F : |a| \leq \|x^*\|\}$. Just verify that B_{X^*} is closed in this product.)

Given a Banach (or even normed) space X, there is a natural mapping $x \mapsto \hat{x}$ of $X \to C(B_{X^*}, w^*)$ defined by $\hat{x}(x^*) = x^*(x)$ ($x \in$

$X, x^* \in B_{X^*}$). Here $C(B_{X^*}, w^*)$ is the space of all scalar-valued continuous functions on the compact Hausdorff space (B_{X^*}, w^*). Of course, one need observe that $\hat{x}$ is w^*-continuous on B_{X^*} for $x \in X$, but this is immediate, since the w^* topology is a relativization to X^* of the product topology on F^X and $\hat{x}$ acts on F^X as a "coordinate projection" or pointwise evaluation. One should also realize that the map $\hat{}$ is linear and, by the Hahn-Banach theorem, norm-preserving when $C(B_{X^*}, w^*)$ is given the supremum norm. *That is, X is isometrically isomorphic to a subspace of $C(B_{X^*}, w^*)$.*

This preliminary observation is valid for any normed space X. We want to see that, when X is separable, $C(B_{X^*}, w^*)$ (and hence also X) is isometrically isomorphic to a subspace of $C[0, 1]$. Since $C[0, 1]$ is separable, so must be $C(B_{X^*}, w^*)$. The reader may know that $C(K)$ (K compact Hausdorff) is separable if and only if K is metrizable; this suggests

PROBLEM 2.1 (i): *If X is separable, then (B_{X^*}, w^*) is metrizable.*

Actually, one can, upon occasion, write down a specific metric on B_{X^*} which determines the w^* topology. Suppose one has a bounded sequence $(x_n)_{n=1}^{\infty}$ in X whose linear span is dense in X. Define $d(\ ,\ )$ on X^* by $d(x^*, y^*) = \sum_{n=1}^{\infty} 2^{-n} |x^*(x_n) - y^*(x_n)|$. Then d is a metric on X^*, which generates the w^* topology *on B_{X^*}*. (The metric d does not generate the w^* topology on X^* itself except when X is finite dimensional.) □

It will be useful to have a sufficient condition on compact Hausdorff spaces K and L which guarantees that $C(L)$ imbeds into $C(K)$.

PROBLEM 2.1 (ii): *If there is a continuous function from the compact Hausdorff space K onto the compact Hausdorff space L, then $C(L)$ is isometrically isomorphic to a subspace of $C(K)$.*

There are now two ways to complete Problem 2.1. The more direct way is to show that, when X is separable, there is a continuous function from $[0, 1]$ onto the compact metrizable space (B_{X^*}, w^*). This can be done by proving that there is a continuous function from $[0, 1]$ onto any compact, connected, locally connected metric space, and then showing that (B_{X^*}, w^*) is connected and locally connected. Indeed, a metric of the type mentioned after the statement of Problem 2.1(i), which generates the w^* topology on B_{X^*} has the property that all ϵ-balls are convex; one can use this to show that (B_{X^*}, w^*) is even arcwise and locally arcwise connected.

The second way to complete the proof of Problem 2.1 uses the Cantor set, which we will denote by Δ. We regard Δ as the closed subset of $[0, 1]$ which one gets by deleting "middle third" intervals, but recall that the Cantor set is homeomorphic to the countable product $(\{0, 1\}^{\mathbb{N}}, \tilde{\beta})$, when the two point space $\{0, 1\}$ is given the discrete topology β and $\tilde{\beta}$ is the product topology $\beta^{\mathbb{N}}$.

Using the representation of Δ as a subset of $[0, 1]$, it can be shown that:

PROBLEM 2.1 (iii): $C(\Delta)$ *is isometrically isomorphic to a subspace of* $C[0, 1]$.

Of course, Δ is not a continuous image of $[0, 1]$, (since $[0, 1]$ is connected and Δ is not connected), so Problem 2.1 (iii) does not follow from Problem 1 (ii).

Given Problem 2.1 (i)–(iii), Problem 2.1 can be completed by showing:

PROBLEM 2.1 (iv): *Every compact metric space is the continuous image of* Δ.

One way of solving Problem 2.1 (iv) is hinted at in Exercise O, p. 165 of [10]. Another approach is to represent the compact metric space K as a space of 0's and 1's (i.e., as a subset of $\{0, 1\}^{\mathbb{N}}$) as follows: K has a countable base, say $(O_n)_{n=1}^{\infty}$, of open sets, since it is a separable metric space. Correspond the point k in K to the point $g_k \in \{0, 1\}^{\mathbb{N}}$ by $g_k(n) = 1$ if $k \in O_n$, and $g_k(n) = 0$ if $k \notin O_n$. Let $G = \{g_k : k \in K\}$. The mapping $k \mapsto g_k$ is certainly not a homeomorphism from K onto G when G is considered as a subset of the Cantor set. However, if τ is the (non-Hausdorff!) topology $\{\varnothing, \{1\}, \{0, 1\}\}$ on $\{0, 1\}$ and $\tilde{\tau}$ is topology of the product space $(\{0, 1\}^{\mathbb{N}}, \tau^{\mathbb{N}})$, then the correspondence $k \to g_k$ is a homeomorphism from K onto $(G, \tilde{\tau})$.

It is clear that the identity mapping i: $(\{0,1\}^{\mathbb{N}}, \tilde{\beta}) \to (\{0,1\}^{\mathbb{N}}, \tilde{\tau})$ is continuous; hence, so is its restriction i: $(G, \tilde{\beta}) \to (G, \tilde{\tau})$. Thus we have a continuous mapping from a subset $(G, \tilde{\beta})$ of the Cantor set onto $(G, \tilde{\tau})$, which is homeomorphic to K. Unfortunately, G is generally not closed in the Cantor space $(\{0,1\}^{\mathbb{N}}, \tilde{\beta})$, so we need to extend i: $(G, \tilde{\beta}) \to (G, \tilde{\tau})$ to a continuous mapping $\bar{i}$: $(\bar{G}, \tilde{\beta}) \to (G, \tilde{\tau})$, where $\bar{G}$ is the $\tilde{\beta}$-closure of G in the Cantor space $(\{0,1\}^{\mathbb{N}}, \tilde{\beta})$. One can show that if L_1 and L_2 are disjoint $\tilde{\tau}$ closed sets in G, then $\bar{L}_1$ and $\bar{L}_2$ (the bar $^-$ again means $\tilde{\beta}$-closure in $\{0,1\}^{\mathbb{N}}$) are disjoint. Using this it can be shown that i can be extended as follows: given $g \in \bar{G}$, let U_n be a sequence of $\tilde{\beta}$ open neighborhoods of g whose intersection is $\{g\}$. The set $\bigcap_{n=1}^{\infty} \text{cl}[i(U_n \cap G)]$ is a singleton, whose only element we denote by $\bar{i}(g)$. The mapping $g \to \bar{i}(g)$ is a continuous mapping from $(\bar{G}, \tilde{\beta})$ onto $(G, \tilde{\tau})$.

Finally, one completes the proof of Problem 2.1 (iv) by showing that every closed subset of Δ is a retract of Δ; i.e., if $C \subseteq \Delta$ is closed, then there is a continuous mapping from Δ to C which is the identity on C. (This is most easily done by regarding Δ as a subset of $[0,1]$.)

* * *

Before discussing complementably universal Banach spaces, we need to mention some facts concerning bases in Banach spaces. The standard terminology is that a sequence $(x_n)_{n=1}^{\infty}$ in a Banach space X is called *basic* provided that each vector $x \in \overline{\text{span}}\,(x_n)$ (the closure in X of the linear span of (x_n)) has a *unique* representation $x = \sum_{n=1}^{\infty} \alpha_n x_n$; i.e., there is a unique sequence of scalars (α_n) which satisfies $\|x - \sum_{n=1}^{N} \alpha_n x_n\| \to 0$ as $N \to \infty$. One can then define functionals x_n^* on the subspace $\overline{\text{span}}(x_n)$ by $x_n^*(x) = \alpha_n$, where $x = \sum_{i=1}^{\infty} \alpha_i x_i$. The uniqueness condition guarantees that the functionals (x_n^*) are linear and satisfy the *biorthogonality condition*:

$$x_n^*(x_m) = \delta_{nm} = 1 \quad \text{for } n = m,$$
$$= 0 \quad \text{for } n \neq m.$$

A non-obvious fact (proved by Banach) is that the x_n^* are always continuous. Rather than go into that here (the interested reader can consult, e.g., p. 207 of [22], for a proof), let us simply take the continuity of the biorthogonal functionals x_n^* as part of the definition of basic sequence. So, for us, a basic sequence is a sequence (x_n) for which there is a sequence $(x_n^*) \subseteq \overline{(\text{span}(x_n))}^*$ which is biorthogonal to (x_n) such that $x = \sum_{n=1}^\infty x_n^*(x) x_n$ for each $x \in \overline{\text{span}}(x_n)$. If $\overline{\text{span}}(x_n) = X$, we say that (x_n) is a *basis for X*.

Given a basis (x_n) for X with biorthogonal functionals $(x_n^*) \subseteq X^*$, define the partial sum operators $S_n: X \to X$ by

$$S_n x = \sum_{i=1}^n x_i^*(x) x_i.$$

Thus the S_n are finite rank operators (continuous and linear, since the x_n^*'s are) on X which satisfy $S_n x \to x$ for all $x \in X$. Therefore a Banach space with a basis has the b.a.p.

We note several properties of the partial sum operators:

$$S_n x \to x \quad \text{for all } x \in X; \tag{2.1}$$

$$S_n S_m = S_{\min(n,m)} \quad \text{for all } n, m = 1, 2, \ldots; \tag{2.2}$$

$$\dim S_n X = n; \tag{2.3}$$

$$\sup \|S_n\| = \lambda < \infty. \tag{2.4}$$

(Relation (2.4) follows from (2.1) via the uniform boundedness principle. The number λ is called the *basis constant* of the basis.)

Given any sequence (S_n) of operators X which satisfy (2.1)–(2.3) (and hence also (2.4)), and picking $x_n \neq 0$ such that elements $x_n \in (S_n - S_{n-1})X$ (where $S_0 \equiv 0$), we have that (x_n) is a basis for X whose partial sum operators are S_n. Note also that the basis constant λ is the smallest constant for which the inequality

$$\left\| \sum_{i=1}^n \alpha_i x_i \right\| \leq \lambda \left\| \sum_{i=1}^m \alpha_i x_i \right\| \tag{2.4a}$$

holds for all finite sequences (α_i) of scalars and all positive integers $n \leq m$. Indeed the smallest such λ is just the supremum over n of the norm of S_n restricted to span (x_n), and $\|S_n\|$ is equal to the norm of S_n restricted to any dense subspace. (The reader should also observe that if (x_i) is a sequence of non-zero vectors which satisfies (2.4a), then (x_i) is basic.)

Let us mention the notion of equivalence of basis. Bases (x_n) and (y_n) for Banach spaces X and Y, respectively, are said to be *equivalent* provided the map $x_n \mapsto y_n$ extends to an isomorphism from X onto Y. This implies that there are constants $0 < a, b < \infty$ such that

$$a \left\| \sum_{i=1}^{n} a_i x_i \right\| \leq \left\| \sum_{i=1}^{n} a_i y_i \right\| \leq b \left\| \sum_{i=1}^{n} a_i x_i \right\| \qquad (2.5)$$

is satisfied for every finite sequence $(a_i)_{i=1}^{n}$ of scalars. Conversely, if (2.5) is satisfied, then we can define an isomorphism ($\equiv$*linear homeomorphism*) S: span $(x_n) \to$ span (y_n) by

$$S\left(\sum_{i=1}^{n} a_i x_i \right) = \sum_{i=1}^{n} a_i y_i.$$

The operator S extends to an isomorphism from $X = \overline{\text{span}}(x_n)$ onto $Y = \overline{\text{span}}(y_n)$ and $Sx = \sum_{n=1}^{\infty} x_n^*(x) y_n$, where (x_n^*) is the sequence in X^* which is biorthogonal to (x_n).

With these preliminaries out of the way, we can state a recent (slight) improvement by Schechtman [21] of a result of Pełczyński [16].

PROBLEM 2.2: *There is a basis (e_n) for some Banach space U which is "complementably universal" for the class of all bases. That is, given any Banach space X with a basis (x_n), there is a subsequence (e_{n_i}) of (e_n) such that (e_{n_i}) is equivalent to (x_i). Further, there is a projection P from U onto $\overline{\text{span}}(e_{n_i})$. If $(e_n^*) \subseteq U^*$ is biorthogonal to (e_n), then P is defined by*

$$Pu = \sum_{i=1}^{\infty} e_{n_i}^*(u) e_{n_i};$$

that is, P is the projection onto $\overline{\text{span}}(e_{n_i})$ whose kernel is $\overline{\text{span}}\{e_j : j \notin (n_i)\}$.

Before hints on how to solve Problem 2.2 are given, two simple remarks are in order. First, a subsequence of a basic sequence is basic (the uniqueness condition follows from the continuity of the biorthogonal functionals). Second, it is *not* true that, given a basis (x_n) for X and subsequence (x_{n_i}) of (x_n), the subspace $\overline{\text{span}}(x_{n_i})$

must be complemented in X. For example, there are bases (x_n) for $C[0,1]$ which have a subsequence (x_{n_i}) for which $\overline{\operatorname{span}}(x_{n_i})$ is reflexive and therefore (by the remarks in the introduction) not complemented in $C[0,1]$.

Schechtman's proof of Problem 2.2 uses Problem 2.1. Since every separable Banach space is isometrically isomorphic to a subspace of $C[0,1]$, we only have to look at basic sequences in the one fixed space $C[0,1]$. Of course, $C[0,1]$ is uncountable, so this is a lot of sequences; however, we can restrict attention to basic subsequences of one fixed dense sequence in $C[0,1]$.

PROBLEM 2.2 (i) (Principle of small perturbations): *Let (x_n) be a basic sequence in a Banach space X. Then there is a sequence (ϵ_n) of positive numbers such that if $y_n \in X$ and $\|y_n - x_n\| < \epsilon_n$ for $1 \leq n < \infty$, then (y_n) is a basic sequence which is equivalent to (x_n).*

Let (z_n) be a fixed sequence of non-zero vectors in $C[0,1]$ which is dense in $C[0,1]$. It follows from Problems 2.1 and 2.2 (i) that every basic sequence is equivalent to a basic subsequence of (z_n), but of course (z_n) is not itself basic. Suppose, however, that we let $(\tilde{z}_i)$ represent a linearly independent sequence in some vector space (e.g., $\tilde{z}_i$ could be the ith unit vector in the space of all scalar-valued sequences) and define a norm $\|\|\ \|\|$ on span $(\tilde{z}_i)$ by $\|\|\sum_{i=1}^n a_i \tilde{z}_i\|\| = \sup_{1 \leq m \leq n} \|\sum_{i=1}^m a_i z_i\|$. Let $(Z, \|\|\ \|\|)$ be the completion of span $(\tilde{z}_i)$ under the norm $\|\|\ \|\|$. Then it is easy to check that $(\tilde{z}_i)$ is a basis for Z with basis constant 1. Suppose now that (z_{n_k}) is a basic subsequence of (z_n) and λ is the basis constant for (z_{n_k}). Then for any finite sequence $(\alpha_i)_{i=n}^n$ of scalars,

$$\left\|\sum_{i=1}^n \alpha_i z_{n_i}\right\| \leq \left\|\left\|\sum_{i=1}^n \alpha_i \tilde{z}_{n_i}\right\|\right\| \leq \lambda \left\|\sum_{i=1}^n \alpha_i z_{n_i}\right\|,$$

so that $(\tilde{z}_{n_i})$ is equivalent to (z_{n_i}). Thus $(\tilde{z}_i)$ is a "universal basic sequence." However, there is no reason to suspect that $\overline{\operatorname{span}}(\tilde{z}_{n_i})$ is complemented in Z, so the constructed space Z does not solve Problem 2.2.

Schechtman's construction of the complementably universal space U uses a modified version of Z. Let C be the set of all finite length increasing sequences of natural numbers; and for $c = (i_1 < i_2 < \cdots < i_k) \in C$, let $l(c) = i_k$ be the last term of c. It is convenient to order the desired universal basis for U by the countable set C rather than by the natural numbers $\mathbb{N}$ (but of course we will later have to put a sequential order onto the "basis" to make it into basic sequence).

So let $(e(c))_{c \in C}$ be a linearly independent sequence in some vector space. We will construct a norm $|\ |$ on $\operatorname{span}(e(c))_{c \in C}$ and let U be the completion of this space. The space U will have the property that if $1 \leq n_1 < n_2 < \cdots$ is a sequence of integers, then $e(n_1), e(n_1, n_2), e(n_1, n_2, n_3), \ldots$ is a basic sequence in U which is equivalent to the sequence (z_{n_i}) in Z, and the natural projection from U onto $\overline{\operatorname{span}}(e(n_1), e(n_1, n_2), \ldots)$ is bounded. Let D be the collection of all finite sequences $(c_1, c_2, \ldots, c_k)$ in C such that c_i is an initial sequence of c_{i+1} and length $c_i = i$ $(1 \leq i \leq k)$. That is, a typical element of D is of the form $\{(i_1), (i_1, i_2), \ldots, (i_1, i_2, \ldots, i_k)\}$, where $1 \leq i_1 < i_2 < \cdots < i_k$ is a finite sequence of integers. Now define $|\ |$ on $\operatorname{span}(e(c))_{c \in C}$ by

$$\left| \sum a_c e(c) \right| = \sup \left\{ \left\| \sum_{i=1}^{k} a_{c_i} \tilde{z}_{l(c_i)} \right\| : (c_1, \ldots, c_k) \in D \right\}.$$

From the construction it is clear that if $1 \leq i_1 < i_2 < \cdots$, and $c_j = (i_1, i_2, \ldots, i_j)$ for $1 \leq j < \infty$, then $\|\sum \alpha_j \tilde{z}_{i_j}\| = |\sum \alpha_j e(c_j)|$ and hence in the completion U of $(\operatorname{span} e(c), |\ |)$, the sequence $(e(c_j))$ is a basic sequence which is equivalent to $(\tilde{z}_{i_j})$. Further, it is also easy to see from the definition of $|\ |$ that $|\sum \alpha_j e(c_j)| \leq |\sum \alpha_j e(c_j) + \sum_{c \in L} \alpha_c e(c)|$ for any finite set $L \subseteq C$ for which $L \cap (c_j)_{j=1}^{\infty} = \emptyset$. This means that there is a norm one projection from U onto $\overline{\operatorname{span}}(e(c_j))$ whose kernel is $\overline{\operatorname{span}}\{e(c) : c \in C, c \notin (c_j)_{j=1}^{\infty}\}$.

As mentioned above, we need to order the set $(e(c))_{c \in C}$ into a sequence so as to make it basic. This is easy; just take any one-to-one mapping ϕ from C onto $\mathbb{N}$ such that $\phi(c_1) < \phi(c_2)$ whenever c_1 is an initial segment of c_2 and order $(e(c))$ as $e(\phi^{-1}(1)), e(\phi^{-1}(2)), e(\phi^{-1}(3)), \ldots$. One can verify that for all scalars (α_i) and $n < m$,

$$\left| \sum_{i=1}^{n} \alpha_i e(\phi^{-1}(i)) \right| \leq \left| \sum_{i=1}^{m} \alpha_i e(\phi^{-1}(i)) \right|. \quad \square$$

Our next goal is to see that every separable Banach space which has the b.a.p. is complemented in some space with a basis, and hence by Problem 2.2 is complemented in Pełczyński's space U. We need a type of approximation structure that is intermediate between b.a.p. and basis. Suppose (S_n) is a sequence of operators on a Banach space X which satisfies (1), (2), (4), and

$$0 < \dim S_n X < \dim S_{n+1} X < \infty; \tag{3'}$$

that is, (S_n) acts like the partial sum projections with respect to some basis, except that $\dim S_n X$ is not necessarily n. One can, however, define the basis-like structure $E_n = (S_n - S_{n-1})X$ (where $S_0 = 0$). Then (E_n) is a sequence of finite dimensional subspaces of X, $E_n \cap E_m = \varnothing$ for $n \neq m$, and for each $x \in X$, there is a unique sequence $e_n \in E_n$ (namely, $e_n = (S_n - S_{n-1})x$) which satisfies $\|x - \sum_{n=1}^N e_n\| \to \infty$ as $N \to \infty$. For obvious reasons, (E_n) is called the *finite dimensional decomposition* (or f.d.d.) for X determined by S_n. As in the basis case, if (E_n) is a sequence of finite dimensional subspaces of X and there is a constant $\lambda < \infty$ such that $\|\sum_{i=1}^m e_i\| \leq \lambda \|\sum_{i=1}^m e_i\|$ whenever $e_i \in E_i$, $n < m$, then (E_n) forms an f.d.d. for $\operatorname{span} E_n$.

PROBLEM 2.3: *If X has the b.a.p., then X is isomorphic to a complemented subspace of a Banach which has an f.d.d.*

Hint: Suppose (T_n) is a sequence of finite rank operators on X which satisfies $x = \sum T_n x$ for each $x \in X$. Let $E_n = T_n X$ and define Z to be the space of all sequences (z_n) for which $z_n \in E_n$ and $\sum_{n=1}^\infty z_n$ converges in X, with the norm $\||(z_n)\||_Z = \sup_N \|\sum_{n=1}^N z_n\|_X$. Then $T: X \to Z$ defined by $Tx = (T_n x)_{n=1}^\infty$ defines an isomorphism from X onto a complemented subspace of Z. □

It is not known whether every Banach space which has an f.d.d. also has a basis (in nuclear Fréchet spaces, this is false [14]). The obvious approach of picking out a basis from each summand of an f.d.d. does not work, because it is not known whether there is a constant $\lambda < \infty$ such that every finite dimensional space has a basis whose basis constant is $\leq \lambda$. However,

PROBLEM 2.4: *If* $\dim E < \infty$, *then there exist a space F with* $\dim F < \infty$ *and an operator* $T: E \to F$ *with* $\|T\| \leq 2$, $\|T^{-1}\| \leq 1$ *for which TE is the range of a projection on F whose norm is* ≤ 2.

To prove Problem 2.4 pick a basis $(e_i)_{i=1}^n$ in E with biorthogonal functionals $(e_i^*)_{i=1}^n \in E^*$, normalized so that $\|e_i\| = 1$, and let $M = \max_{1 \leq i \leq n} \|e_i^*\|$, and choose an integer m so that $n^{m-1} \geq M$. (Actually, it is known that (e_i) can be chosen so that also $\|e_i^*\| = 1$.) Let $T_i: E \to E$ be the one dimensional operator $T_i x = e_i^*(x) e_i$. Of course, $\sum_{i=1}^n T_i$ is the identity operator I on E, but the obvious estimate on the partial sum operators $S_k = \sum_{i=1}^k T_i$ gives only $\|S_k\| \leq kM \leq n^m$. However, if one sets $\tilde{T}_{nk+i} = n^{-m} T_i$ for $0 \leq k \leq n^m - 1$, then $\sum_{i=1}^{n^{m+1}} \tilde{T}_i = I$ and $\max_{1 \leq N \leq n^{m+1}} \|\sum_{i=1}^N \tilde{T}_i\| \leq 2$. Let $F = \{(x_i)_{i=1}^{n^{m+1}} : x_i \in \tilde{T}_i E\}$ with the norm

$$\|\|(x_i)_{i=1}^{n^{m+1}}\|\|_F = \max_{1 \leq N \leq n^{m+1}} \|\sum_{i=1}^N x_i\|_E.$$

Check that the operator $T: E \to F$ defined by $Tx = (\tilde{T}_i x)_{i=1}^{n^{m+1}}$ has the desired properties.

PROBLEM 2.5: *Every Banach space with an f.d.d. is isomorphic to a complemented subspace of a Banach space which has a basis.*

Hint: If (E_n) is an f.d.d. for X, then there are finite dimensional spaces (F_n) such that $E_n \oplus F_n$ (normed, for example, by $\|e \oplus f\| = (\|e\|^2 + \|f\|^2)^{1/2}$) has a basis with basis constant ≤ 4. Show that $X \oplus (\sum F_n)_{l_2}$ has a basis. Here $(\sum F_n)_{l_2} = \{(f_n) : f_n \in F_n$ and $\|\|(f_n)\|\| \equiv (\sum \|f_n\|^2)^{1/2} < \infty\}$. □

There are separable Banach spaces isomorphically different from U which are complementably universal for all separable Banach spaces which possess the b.a.p. (e.g., $U \oplus Y$, where Y is any separable space which fails the a.p.). However, Pełczyński's decomposition method [15] allows one to prove:

PROBLEM 2.6: *If Z is a separable Banach space which has the b.a.p. and U is isomorphic to a complemented subspace of Z, then U and Z are isomorphic.*

This is a Cantor-Bernstein problem; U is complemented in Z, Z is complemented in U, from which we conclude that U and Z are the same up to isomorphism. Actually, Problem 2.6 is a (very) special case of Problem 2.7 below. Before stating Problem 2.7 we mention the concept of an l_p-*sum* of spaces (which was used already in the hint after Problem 2.5). Given a sequence X_n of Banach spaces, $(\Sigma X_n)_p$ or $(X_1 \oplus X_2 \oplus \cdots)_p$ denotes the space of all sequences (x_n) for which $x_n \in X_n$ and $(\Sigma \|x_n\|^p)^{1/p} < \infty$, normed by $\|\|(x_n)\|\| = (\Sigma \|x_n\|^p)^{1/p}$; the space $(\Sigma X_n)_p$ is a Banach space whenever each X_n is. Certain isomorphic identities are easily established. If $Y \sim X$ means Y is isomorphic to X, it is easy to establish, for example, that $(\Sigma X_n)_p \oplus (\Sigma Y_n)_p \sim (\Sigma(X_n \oplus Y_n))_p$ or that $(\Sigma X_n)_p \sim X_1 \oplus (\Sigma_{n=2}^\infty X_n)_p$. Here $X_n \oplus Y_n$ can be endowed with any norm which satisfies

$$a \max(\|x\|, \|y\|) \leq \|(x,y)\| \leq b(\|x\| + \|y\|) \quad (x \in X_n, y \in Y_n),$$

where $0 < a, b < \infty$, and a, b do not depend on n. Another easy fact is that

$$((X \oplus X \oplus \cdots)_p \oplus (X \oplus X \oplus \cdots)_p \oplus \cdots)_p$$

is isometrically isomorphic to $(X \oplus X \oplus \cdots)_p$, no matter what X is.

PROBLEM 2.7: *If $X \sim (X \oplus X \oplus \cdots)_p$ for some p, $1 \leq p < \infty$, Y is a complemented subspace of X, and Y contains a complemented subspace which is isomorphic to X, then X and Y are isomorphic.*

Hint: Write $X \sim Y \oplus Z$, $Y \sim X \oplus W$. Juggle symbols to show that $Y \sim X \oplus Y$ and $X \sim X \oplus Y$. □

To derive Problem 2.6 from Problem 2.7, show first that, for $1 \leq p < \infty$, then $(X \oplus X \oplus \cdots)_p$ has the b.a.p. whenever X does. Thus U contains a complemented copy of $(U \oplus U \oplus \cdots)_p$, for all $1 \leq p < \infty$.

From the discussion above we see that the separable space U is isomorphic to $(U \oplus U \oplus \cdots)_p$ for all $1 \leq p < \infty$. It seems that U is the first example of a separable space with this property.

From Problem 2.6 it is possible to conclude that U has many other properties. For example, in the case of complex scalars, U is isomorphic to a uniform algebra. (Recall that a *uniform algebra* is a closed subalgebra of $C(K)$ for some compact Hausdorff space K. Here $C(K)$ is the space of all continuous scalar-valued functions on K. Of course, multiplication on $C(K)$ is defined pointwise and the norm on $C(K)$ is the supremum norm.) Indeed, this follows from a theorem of Milne's [13] which we state as:

PROBLEM 2.8: *If X is a separable Banach space (complex scalars), then X is isomorphic to a complemented subspace of a separable uniform algebra A.*

Hint: Milne's space A is the closure of the subalgebra $\tilde{A}$ of $C(B_{X^*}, \text{weak}^*)$ which is generated by $\{1\} \cup \hat{X}$. Here $\hat{X} = \{\hat{x} \in C(B_{X^*}) : x \in X\}$ where $\hat{x} : B_{X^*} \to \mathbb{C}$ is defined by $\hat{x}(x^*) = x^*(x)$. A typical element of $\tilde{A}$ is a "polynomial" of the form $f = c_0 + \sum_{i=1}^{n(1)} c_i \hat{x}_i + \sum_{i=1}^{n(2)} d_i \hat{y}_i \hat{z}_i + \cdots$; here c_i and d_i are in $\mathbb{C}$; x_i, y_i, and z_i are in X. If one sets $\tilde{P}f = \sum_{i=1}^{n(1)} c_i \hat{x}_i$, then $f \to \tilde{P}f$ defines a projection from $\tilde{A}$ onto $\hat{X}$.

PROBLEM 2.8 (i): *The operator $\tilde{P}$ is a norm one projection, and hence extends to a norm one projection from A onto $\hat{X}$.*

Hint: Let $f \in \tilde{A}$ have the representation above and assume $\|f\| = \sup\{|f(x^*)| : x^* \in B_{X^*}\} \leq 1$. Fix $x_0^* \in B_{X^*}$ and consider the complex polynomial

$$u(z) = c_0 + \left(\sum_{i=1}^{n(1)} c_i \hat{x}_i(x_0^*)\right) z + \left(\sum_{i=1}^{n(2)} d_i \hat{y}_i(x_0^*) \hat{z}_i(x_0^*)\right) z^2 + \cdots + (\cdots) z^n,$$

where n is the degree of f. Argue that $|u(z)| \leq 1$ whenever $|z| \leq 1$. Use the Cauchy integral formula to show that $|u'(0)| \leq 1$, and observe that $u'(0) = \sum_{i=1}^{n(1)} c_i \hat{x}_i(x_0^*) = \tilde{P}f(x_0^*)$.

PROBLEM 2.8 (ii): *If, in the notation of Problem 2.8, X has the b.a.p., then A may be taken to have the b.a.p.*

Hint: It is enough to prove this for $X = U \equiv$ Pełczyński's universal basis space. But the basis for U yields a sequence of finite rank operators T_n on U for which $\|T_n\| = 1$ and $T_n x \to x$ for all $x \in U$. Show that the generated algebra $A = A(U)$ has this same property.

PROBLEM 2.8 (iii): *The universal basis space U is isomorphic to a uniform algebra.*

PROBLEM 2.9: *Every separable uniform algebra is isometrically algebra isomorphic to a subalgebra of $C(\Delta)$.*

Hint: You've already done this one.

3. THE NON-EXISTENCE OF A SEPARABLE COMPLEMENTABLY UNIVERSAL BANACH SPACE

The reader should accept the following axiom, a stronger version of which is proved in Section 4.

AXIOM: *For each $2 < p < \infty$, there is a subspace E_p of l_p which fails to have the bounded compact approximation property.*

We say that a separable space X has the *bounded compact approximation property* (b.c.a.p., for short) provided that there is a sequence (T_n) of compact operators on X so that $T_n x \to x$ for each $x \in X$. Of course, $\sup \|T_n\| < \infty$ for such a sequence by the uniform boundedness principle. From this it is easy to prove:

PROBLEM 3.1: *Suppose X fails to have the b.c.a.p. Given any $\lambda < \infty$, there are $\epsilon > 0$ and a finite subset F of the unit ball B_X of X so that if T is a compact operator on X and $\max_{f \in F} \|Tf - f\| < \epsilon$, then $\|T\| > \lambda$.*

A result that more or less goes back to Banach is needed:

PROBLEM 3.2: *Let $1 < p < r < \infty$. Then every operator from a subspace of l_r into a subspace of l_p is compact.*

Hint: First observe that if (x_n) is a disjointly supported sequence in l_p and $0 < a = \inf \|x_n\| \leq \sup \|x_n\| = b < \infty$, then $a(\Sigma |\alpha_i|^p)^{1/p} \leq \|\Sigma \alpha_i x_i\| \leq b(\Sigma |\alpha_i|^p)^{1/p}$ for all choices (α_i) of scalars. Thus, (x_i) is equivalent to the natural unit vector basis of l_p. Then use a "gliding hump" procedure and the principle of small perturbations (i.e., Problem 2.2 (i)) to solve:

PROBLEM 3.3: *If* $\inf \|x_n\| > 0$, $x_n \in l_p$, $(1 < p < \infty)$, *and if* (x_n) *converges weakly to zero, then* (x_n) *has a subsequence which is equivalent to the unit vector basis of* l_p.

Now suppose $T: E \to F$ is not compact, where $E \subseteq l_r$, $F \subseteq l_p$. Then since E is reflexive, there is a sequence (x_n) of unit vectors in E which converges weakly to zero, but $\inf \|Tx_n\| > 0$. Since also (Tx_n) converges weakly to 0, we conclude from Problem 3.3 that there is a subsequence (y_n) of (x_n) such that (y_n) is equivalent to the unit vector basis of l_r and (Ty_n) is equivalent to the unit vector basis of l_p. This cannot be, since $p < r$.

PROBLEM 3.4: *There is no separable Banach space which contains, for each* $2 < p < \infty$, *a complemented subspace isomorphic to* E_p.

Suppose the separable space X contains, for each $2 < p < \infty$, a subspace F_p isomorphic to E_p, and suppose Q_p is a projection from X onto F_p. By Problem 3.2, $Q_p Q_r$ is compact when $p < r$. Intuition suggests that, since there are uncountably many F_p in the separable space X, many of them have to be "close together." But if F_p and F_r are "close," then $Q_p Q_r$ is "close" to the identity on F_p. This can be made precise as follows: For some $\lambda < \infty$, then $\|Q_p\| < \lambda$ for uncountably many $p \in (2, \infty)$, say, for $p \in A$. Each space F_p fails to have the b.c.a.p. since each space E_p does, so there are finite sets $(y_i^p)_{i=1}^{n(p)}$ in the unit ball of F_p and a number $\epsilon_p > 0$ such that if T is a compact operator on F_p and $\|T\| < \lambda^2$, then $\|Ty_i - y_i\| > \epsilon_p$ for some i, $1 \leq i \leq n(p)$. One can now pass to an uncountable subset B of A such that $n(p) = n$ is constant for $p \in B$, and $\epsilon_p \geq \epsilon > 0$ for $p \in B$. The finite sets $(y_i^p)_{i=1}^n$ for $p \in B$ lie in the separable metric space X; hence, for some $p < r$, $p, r \in B$, we have $\max_{1 \leq i \leq n} \|y_i^p - y_i^r\| < \epsilon(\lambda + \lambda^2)^{-1}$. One can check that the restriction (call it T) of

$Q_p Q_r$ to F_p is a compact operator on F_p which satisfies $\|T\| \leq \lambda^2$ but $\max_{1 \leq i \leq n} \|y_i^p - Ty_i^p\| < \epsilon$. This contradiction completes the sketch of the proof of Problem 3.4.

REMARK: In [8] two extensions of Problem 3.4 are presented:

3.5: *There is no separable Banach space which is complementably universal for the class of separable Banach spaces which possess the approximation property.*

3.6: *Fix $2 < p < \infty$. There is no separable Banach space which is complementably universal for the class of subspaces of l_p.*

Prior to the Enflo-Davie-Figiel examples of subspaces of l_p ($2 < p < \infty$) which fail to have the a.p., there was not known to exist an infinite dimensional subspace of l_p which is not isomorphic to l_p. Now we know from 3.6 that there are, for each fixed $2 < p < \infty$, uncountably many mutually non-isomorphic subspaces of l_p.

4. COUNTEREXAMPLES TO THE APPROXIMATION PROBLEM

A bounded, biorthogonal sequence for a normed space is a sequence of pairs $(x_n, x_n^*)_{n=1}^\infty$ with $x_n \in X$, $x_n^* \in X^*$, which satisfies the *boundedness condition*:

$$\sup \|x_n\| \|x_n^*\| < \infty$$

and the *biorthogonality condition*:

$$x_n^*(x_m) = 1 \quad \text{for } n = m,$$
$$= 0 \quad \text{for } n \neq m.$$

Enflo's approach to the approximation problem was to construct a bounded biorthogonal sequence (x_n, x_n^*) such that if T is a finite rank operator on $X = \overline{\text{span}}(x_n)$, and T is "close" to the identity on $\text{span}(x_i)_{i=1}^n$, then either $\|T\|$ is "big" (say, $\geq \lambda(n)$, where $\lambda(n) \to \infty$ as $n \to \infty$), or T is "close" to the identity on $\text{span}(x_i)_{i=1}^{n+1}$. Intuitively, it seems that such a space X should fail to have the b.a.p.

(Enflo's original proof that there is a space which fails to have the a.p. was that there is a reflexive space which fails to have the b.a.p. It then followed from an "abstract nonsense" argument due to Grothendieck that Enflo's space fails to have the a.p.) Of course, there is difficulty in deciding what is the appropriate notion of "closeness to the identity." Enflo discovered that the relevant measure is that the *mean trace* of T over $(x_i)_{i=1}^n$ is close to 1; i.e., that $(1/n)\Sigma_{i=1}^n x_i^*(Tx_i) \approx 1$. (It can be shown that if $\|x_i\| \|x_i^*\|$ and $\|T\|$ both equal 1, and the mean trace $(1/n)\Sigma_{i=1}^n x_i^*(Tx_i)$ of T over $(x_i)_{i=1}^n$ is 1, then, in fact, T is the identity on span $(x_i)_{i=1}^n$.)

What Enflo actually did was to construct a sequence of *finite* biorthogonal systems $(x_n^k, x_n^{k*})_{n=1}^{m(k)}$ in normed spaces $X_k =$ span$(x_n^k)_{n=1}^{m(k)}$ with $\sup_{n,k}\|x_n^k\| \|x_n^{k*}\| < \infty$ and such that if T is an operator on X_k with $\|T\| \leq \lambda(k)$ (where $\lambda(k) \to \infty$ as $k \to \infty$) and the mean trace of T over a certain subset $(x_n^k)_{n=1}^{p(k)}$ of $(x_n^k)_{n=1}^{m(k)}$ is close to 1, then the mean trace of T over $(x_n^k)_{n=1}^{m(k)}$ is also close to 1. The construction was done in such a way that $m(k) \to \infty$ and also $m(k) - p(k) \to \infty$ as $k \to \infty$, so that, in some sense, the finite dimensional spaces X_k have "worse" approximation properties as k increases. Enflo was able to piece together the spaces X_k in an ingenious way to produce a space which fails to have the b.a.p.

Later we will return to Enflo's mean trace concept and his piecing-together construction, as developed and refined by Figiel and Pełczyński [5], Figiel [3], and Davie [1]; but first let us take the more abstract viewpoint of Figiel to constructing a counterexample.

Suppose that X fails to have the c.a.p. Let $L(X)$ be the space of all bounded linear operators on X, and let τ be the topology on $L(X)$ of uniform convergence on compact subsets of X. Thus, a net (T_α) in $L(X)$ converges to T in $L(X)$ provided that, for each compact set $K \subseteq X$, $\lim_\alpha \sup_{k \in K} \|T_\alpha k - Tk\| = 0$. Note that X has the c.a.p. only if the identity operator I on X is in the τ-closure of the compact operators on X. Now τ makes X into a locally convex linear topological space, so that if X fails to have the c.a.p., then there is a τ-continuous linear functional ϕ on $L(X)$ for which $\phi(I) = 1$ and $\phi(T) = 0$ for each compact operator T on X.

This discussion suggests our first problem; however, we should point out that the reader does not need to know anything about locally convex spaces or the topology τ mentioned in the preceding paragraph to work the (very easy):

PROBLEM 4.1: *The Banach space X fails to have the c.a.p. provided that there is a linear functional ϕ on $L(X)$ and a compact set $K \subseteq X$ such that $\phi(I) = 1$, $\phi(T) = 0$ for all compact operators T on X, and there is $\lambda < \infty$ so that $|\phi(T)| \leq \lambda \sup\{\|Tx\| : x \in K\}$ for every $T \in L(X)$.*

Figiel constructed the desired functional ϕ in Problem 4.1 by taking a limit of an appropriate sequence of functionals ϕ_n.

PROBLEM 4.2: *Let (e_n) be a bounded sequence in X and (e_n^*) a bounded sequence in X^*. Suppose that (A_n) is a pairwise disjoint sequence of sets of positive integers with cardinality $A_n = c_n$. Let $\phi_n(T) = c_n^{-1} \sum_{j \in A_n} e_j^*(Te_j)$ for $T \in L(X)$. If either e_n converges weakly to 0, or e_n^* converges weak* to zero, then $\phi_n(T) \to 0$ for each compact operator T on X.*

Eventually we will construct a biorthogonal sequence (e_n, e_n^*) for a subspace of l_p ($1 < p < \infty$) to which we can apply Problems 4.1 and 4.2 to conclude that $X = \overline{\text{span}} \, (e_n)$ fails to have the c.a.p. It is convenient to represent l_p as $(l_p^{n(1)} \oplus l_p^{n(2)} \oplus \cdots)_p$, where each $n(i)$ is of the form 2^n. We will view $l_p^{2^n}$ as the space of scalar-valued functions on the group $\{-1, 1\}^n$, and use some elementary properties concerning characters on this group, so we would like to recall some simple facts from finite group theory.

Given a set G, let $|G|$ denote the cardinality of G. Recall that a *character* on a finite Abelian group G is a homomorphism from G into the unit circle (i.e., a function h from G into the unit circle satisfying $h(gg') = h(g)h(g')$ for all g, g' in G). The set of characters on G is denoted by $W(G)$, or just W if G is understood. (The letter W is used because the characters on $\{-1, 1\}^n$ are often called the *Walsh functions*.)

PROBLEM 4.3: *If G is a finite Abelian group, then $|W| = |G|$.*

Given a finite Abelian group G, let $Y = Y(G)$ be the vector space (with complex scalars) of all complex-valued functions on G.

(If all the characters on G are real-valued, Y can be the space of real-valued functions on G, with real scalars.) When Y is given the norm $\|f\|_p = (\Sigma_{g \in G} |f(g)|^p)^{1/p}$, we denote $(Y, \| \ \|_p)$ by $l_p(G)$. Of course there is an inner product on Y which generates the norm $\| \ \|_2$ on Y; namely, $\langle f, h \rangle = \Sigma_{g \in G} f(g) \overline{h(g)}$, for $f, h \in Y$.

PROBLEM 4.4: *The characters on G are orthogonal in $l_2(G)$ and form a basis for Y.*

As mentioned above, we are interested in groups of the form $\{-1, 1\}^n$. Of course, multiplication in $\{-1, 1\}^n$ is defined coordinatewise: for $g = (a_1, a_2, \ldots, a_n)$, $g' = (b_1, b_2, \ldots, b_n)$, $a_i, b_i = \pm 1$, define $gg' = (a_1 b_1, a_2 b_2, \ldots, a_n b_n)$. The obvious characters on $\{-1, 1\}^n$ are the coordinate projections R_i, defined by $R_i g = a_i$ for $g = (a_1, \ldots, a_n) \in \{-1, 1\}^n$. (The letter R is used because the R_i are often called the *Rademacher functions*.)

PROBLEM 4.5: *Each character on $\{-1, 1\}^n$ is of the form $\Pi_{i \in S} R_i$ for some subset S of $\{1, 2, \ldots, n\}$.*

Here $\Pi_{i \in \varnothing} R_i$ is the constant 1 function. In particular, the characters on $\{-1, 1\}^n$ are all $-1, 1$ valued, so that the real space $l_p(\{-1, 1\}^n)$ has the characters as a basis.

The main fact we need to know about $l_p(G)$ is given by:

PROBLEM 4.6: *Let $G = \{-1, 1\}^n$. Then $W = W(G)$ can be decomposed into two disjoint subsets W^+ and W^-, each of which has $2^{n-1} = \frac{1}{2}|G|$ elements, such that if $y \in Y$ is defined by*

$$y = \sum_{w \in W^+} w - \sum_{w \in W^-} w,$$

then for each $g \in G$, we have $|y(g)| \leq 2^{1+[n/2]}$. (Here $[x]$ is the greatest integer which does not exceed x.)

Later (after Problem 4.8 below), we give a hint for Problem 4.6, but first we want to indicate how Problem 4.6 is related to Enflo's original intuition. The counterexample in l_p to the c.a.p. requires estimates in the norm of $l_p(\{-1, 1\}^n)$ which come out of Problem 4.6. Estimates in l_p ($2 < p < \infty$) can be gotten from estimates in l_2

and l_∞ by means of Hölder's inequality, so attention can be restricted to $l_\infty\{-1,1\}^n$ estimates. (Recall that the $l_\infty(G)$ norm is defined by $\|f\|_\infty = \max_{g \in G} |f(g)|$.) Now let n be fixed, $G = \{-1,1\}^n$, W^+, W^- be as given from Problem 4.6. Note that if e is the identity of G, then $w(e) = 1$ for each character w, so that $\|\Sigma_{w \in W^+} w\|_\infty = 2^{n-1} = \|\Sigma_{w \in W^-} w\|_\infty$. The fact that this is much larger than $\|\Sigma_{w \in W^+} w - \Sigma_{w \in W^-} w\|_\infty \leq 2^{1+[n/2]}$ will imply that $l_\infty(G)$ has "bad" approximation properties in Enflo's sense.

Suppose that $T \in L(Y)$ and T has "small" norm as an operator on $l_\infty(G)$. Assume that T is close to the identity on span$\{w : w \in W^+\}$ in the sense that the mean trace

$$|W^+|^{-1} \sum_{w \in W^+} \langle Tw, \|w\|_2^{-2} w \rangle$$

is about one. (Note that, since W is orthogonal in $l_2(G)$, the biorthogonal functionals to W are defined by $w^*(\cdot) = \langle \cdot, \|w\|_2^{-2} w \rangle$. Also, observe that since $|w(g)| = 1$ for $g \in G$ and $|G| = 2^n$, we have that $\|w\|_2 = 2^{n/2}$.) At first also assume that T is diagonal with respect to W, say $Tw = \lambda_w w$ for $w \in W$ (so that $Tf = \Sigma_{w \in W} \lambda_w \langle f, 2^{-n} w \rangle w$ for each $f \in Y$). The inequality from Problem 4.6:

$$\left\| \sum_{w \in W^+} \lambda_w w - \sum_{w \in W^-} \lambda_w w \right\|_\infty = \left\| T\left(\sum_{w \in W^+} w - \sum_{w \in W^-} w \right) \right\|_\infty$$
$$\leq \|T\| 2^{1+[n/2]},$$

allows us to conclude that the mean trace of T over W^- is about the same as the mean trace of T over W^+, provided that $\|T\|_\infty$ is not too big. Indeed, one can easily check that

$$\left(\sum_{w \in W} w \right)(g) = 2^n \quad \text{if } g = e,$$
$$= 0 \quad \text{otherwise};$$

hence $\|2^{-n} \Sigma_{w \in W} w\|_1 = 1$. But $\|2^{-n} \Sigma_{w \in W} w\|_1$ is the norm in $l_\infty(G)^*$ of the linear functional $\Sigma_{w \in W} w^* = \langle \cdot, \Sigma_{w \in W} 2^{-n} w \rangle$.

Therefore,

$$\left|\sum_{w\in W^+}\lambda_w - \sum_{w\in W^-}\lambda_w\right| = \left|\left(\sum_{w\in W} w^*\right)\left(\sum_{w\in W^+}\lambda_w w - \sum_{w\in W^-}\lambda_w w\right)\right|$$

$$\leq \left\|\sum_{w\in W^+}\lambda_w w - \sum_{w\in W^-}\lambda_w w\right\|_\infty \leq \|T\| 2^{1+[n/2]}.$$

That is: |mean trace of T over W^+ − mean trace of T over W^-| $\leq 2^{-n+1}\|T\| 2^{1+[n/2]} = \|T\| 2^{-[n/2]+2}$ which is quite small if $\|T\| \leq$ constant and n is big.

In the discussion above we assumed that T is diagonal with respect to the basis W. To treat non-diagonal T, we need a few elementary facts concerning the group G acting on Y. We state the next two problems for an arbitrary finite Abelian group G.

Given $g \in G$, we define an operator $U_g \in L(Y)$ by $(U_g f)(g') = f(g^{-1}g')$.

PROBLEM 4.7: *For $w \in W$, then $U_g w = \overline{w(g)} w$. For $w \in W$ and $T \in L(Y)$,*

$$\langle Tw, w\rangle = \sum_{g\in G}(Tw)(g)\overline{w(g)} = \sum_{g\in G}(TU_g w)(g).$$

The operator U_g is an isometric isomorphism on $l_p(G)$ ($1 \leq p \leq \infty$) with inverse $U_{g^{-1}}$.

Given a non-diagonal T on $Y = Y(G)$, the operator T can be averaged against the group $\{U_g : g \in G\}$ to obtain an operator $\tilde{T}$. That is,

$$\tilde{T}f = |G|^{-1}\sum_{g\in G} U_{g^{-1}} T U_g(f).$$

PROBLEM 4.8: $\|\tilde{T}\|_\infty \leq \|T\|_\infty$. *The mean trace of $\tilde{T}$ over any subset A of W is the same as the mean trace of T over A. The operator $\tilde{T}$ commutes with each U_g, and hence $\tilde{T}$ is diagonal with respect to the basis W.*

Hint for Problem 4.8: The point of Problem 4.8 is that $\tilde{T}$ is the diagonal of T relative to the basis W. This purely algebraic fact is perhaps easiest to see by considering T and the U_g to be operators on $l_2(G)$. Note that (by Problem 4.7) $U_{g^{-1}} = U_g^*$, which simply means that $\langle U_{g^{-1}}w, w'\rangle = \langle w, U_g w'\rangle$ for all $w, w' \in W$, $g \in G$. Compute that $\langle U_{g^{-1}} T U_g w, w'\rangle = \overline{w(g)}w'(g)\langle Tw, w'\rangle$, and observe that $\overline{w}(\cdot)w'(\cdot)$ is a character, so that it has average value 0, unless it is constantly 1.

Hint for Problem 4.6: Let $k = [(n-1)/2]$, let $s = n - 2k$, and define $y \in Y(\{-1, 1\}^n)$ by

$$y = \prod_{j=1}^{k} (1 - R_{2j-1} - R_{2j} - R_{2j-1}R_{2j}) \prod_{i=1}^{s} (1 - R_{2k+i}).$$

Recall that $W = \{\prod_{j \in S} R_j : S \subseteq \{1, \ldots, n\}\}$, so if we write $y = \sum_{w \in W} c_w w$, then $c_w = \pm 1$ for each $w \in W$. Argue that $y(e) = 0$ (where e is the identity of $\{-1, 1\}^n$), so that $W^+ = \{w : c_w = 1\}$ and $W^- = \{w : c_w = -1\}$ have the same cardinality. □

We will now attempt to describe the "piecing together" of the spaces $l_p(\{-1, 1\}^n)$ that produces a subspace of l_p that fails to have the c.a.p. For each $j = 1, 2, \ldots, 2^n$ let $G_{n,j}$ be the group $\{-1, 1\}^n$. The characters for $G_{n,j}$ are denoted by $W_{n,j}$ and let $W_{n,j}^+ \cup W_{n,j}^-$ be the decomposition of $G_{n,j}$ given by Problem 4.6. Let Γ be the disjoint union of the $G_{n,j}$ for $n \geq 4$ and $1 \leq j \leq 2^n$, and let V be the corresponding union of all the $W_{n,j}^+$. We represent l_p as $l_p(\Gamma)$ and identify $Y(G_{n,j})$ with the scalar-valued functions on Γ which vanish off $G_{n,j}$. The subspace of $l_p(\Gamma)$ which fails to have the c.a.p. is the closed linear span of a certain sequence $\{e_w\}$, obtained by gluing together elements in $\bigcup_{k=1}^{2^n} W_{n,k}^-$ with elements in $\bigcup_{k=1}^{2^{n-1}} W_{n-1,k}^+$.

Since there are combinatorial difficulties in performing the gluing procedure, we first describe a simple gluing procedure between four disjoint copies $G_{n,1}$, $G_{n,2}$, $G_{n,3}$, and $G_{n,4}$ of G_n which perhaps motivates the actual procedure. Let

$$h: W_{n,2}^- \to W_{n,1}^+, \quad f: W_{n,2}^+ \to W_{n,3}^-, \quad \text{and} \quad b: W_{n,3}^+ \to W_{n,4}^-$$

be one-to-one, onto functions. In $l_p(G_{n,1} \cup G_{n,2} \cup G_{n,3} \cup G_{n,4})$ consider the sets

$$H = \{w + h(w) : w \in W_{n,2}^-\}, \quad F = \{w + f(w) : w \in W_{n,2}^+\},$$
$$\text{and } B = \{w + b(w) : w \in W_{n,3}^+\},$$

and let span $H \cup F \cup B = X$. Of course, $H \cup F \cup B$ is an orthogonal system and thus a basis for X. Suppose that T is an operator on X with small norm which has mean trace over H close to 1. The discussion after Problem 4.6 suggests that the mean trace of T over F should also be close to 1, and, using this, that the mean trace of T over B is close to 1. Of course, we can use infinitely many copies $(G_{n,j})_{j=1}^\infty$ of G_n and extend this gluing together procedure to build an infinite orthogonal sequence M in l_p. However, this sequence M will be basic, so that $\overline{\text{span}\,M}$ has the b.a.p. The trouble with the simple approach is that the degree of closeness of the mean trace of T over $W(G_n)^+$ to the mean trace of T over $W(G_n)^-$ depends on n. This difficulty goes away if n is allowed to increase in the piecing-together construction, but of course this introduces combinatorial difficulties because G_n and G_{n-1} have different cardinalities. What we will do is map an element w in $W_{n,k}^-$ to an element $h(w)$ in $W_{n-1,l}^+$ for some l. This is done in such a way that distinct characters in the same set $W_{n,k}^-$ go into different sets $W_{n-1,l}^+$. This is possible because $|W_{n,k}^-| = 2^{n-1}$ and k varies between 1 and 2^{n-1}. Of course, the cardinality of $\bigcup_{k=1}^{2^n} W_{n,k}^-$ is four times the cardinality of $\bigcup_{k=1}^{2^{n-1}} W_{n-1,k}^+$, so the described mapping h must be at least four-to-one. It is important that $|h^{-1}(w)|$ be bounded independently of w, and, in fact, it can be taken to be 4 for each w. Another important property for h to possess is that $h(W_{n,k}^-)$ and $h(W_{n,j}^-)$ not have big intersection when $k \neq j$; we can actually guarantee that there are at most 2 elements in such intersection. This mapping h is given by:

PROBLEM 4.9: *For each $n = 4, 5, \ldots$ there are two families $\{A_i : i \in I\}$, $\{B_j : j \in J\}$ of mutually pairwise disjoint sets and a function $p : \bigcup_{j \in J} B_j \to \bigcup_{i \in I} A_i$ such that for each $i \in I$ and $j \in J$,*

$$|A_i| = \tfrac{1}{2}|I| = 2^{n-1}, \qquad |B_j| = \tfrac{1}{2}|J| = 2^n,$$
$$|p(B_j) \cap A_i| = 1, \qquad |p^{-1}(a)| = 4$$

for $a \in \bigcup_{i \in I} A_i$, and $|p(B_j) \cap p(B_k)| \leq 2$ when $j, k \in J, j \neq k$. Consequently, there is a function h from $\bigcup \{W_{n,j}^- : n = 5, 6, \ldots; 1 \leq j \leq 2^n\}$ into $V = \bigcup \{W_{n,j}^+ : n = 4, 5, \ldots; 1 \leq j \leq 2^n\}$ such that for each $n \geq 5$, $j, k \leq 2^n$, and $w \in V$, we have:

$$h(W_{n,j}^-) \subseteq \bigcup_{k=1}^{2^{n-1}} W_{n-1,k}^+; \tag{4.9i}$$

$$|h(W_{n,j}^-) \cap W_{n-1,k}^+| = 1, \tag{4.9ii}$$

so that h is one-to-one on each set $W_{n,j}^-$;

$$|h^{-1}(w)| = 4; \tag{4.9iii}$$

$$|h(W_{n,j}^-) \cap h(W_{n,k}^-)| \leq 2 \quad \text{if } j \neq k. \tag{4.9iv}$$

Hint: Let F be a field with 2^{n-1} elements and let D be a four-element subset of $F \setminus \{0\}$. Set

$$I = F \times \{0, 1\} \text{ and } A_i = \{g\} \times F \times \{k\} \text{ for } i = (g, k) \in I;$$

set

$$J = F \times D \text{ and } B_j = I \times \{j\} \text{ for } j \in J;$$

moreover, define

$$p((f, k), (g, d)) = (g + df, (f, k)) \text{ for } ((f, k), (g, d)) \in I \times J = \bigcup_{j \in J} B_j.$$

(Note that $\bigcup A_i$ forms two copies of the two-dimensional plane $F \times F$, and the A_i are vertical lines in these two copies of $F \times F$. The mapping p identifies each B_j with a subset of $\bigcup A_i$ whose intersection with each copy of $F \times F$ is the "same" line which is parallel to one of the directions in D.) □

For $w \in V = \bigcup \{W_{n,j}^+ : n = 4, 5, \ldots; 1 \leq j \leq 2^n\}$, set

$$e_w = w + \sum_{z \in h^{-1}(w)} z,$$

and let X_p be the closed linear span of (e_w) in $l_p(\Gamma)$, where, as before, Γ is the disjoint union of $\{G_{n,j} : n = 4, 5, \ldots,; 1 \leq j \leq 2^n\}$.

Our goal is to show that X_p fails to have the c.a.p. for all $p > 2$. We begin with a trivial problem:

PROBLEM 4.10: *The set $\{e_w : w \in V\}$ is orthogonal in $l_2(\Gamma)$ and hence the biorthogonal functionals e_w^* in any space X_p^* are given by*

$$e_w^*(f) = \langle f, \|e_w\|_2^{-2} e_w \rangle = \|e_w\|_2^{-2} \sum_{\gamma \in \Gamma} f(\gamma) e_w(\gamma).$$

For each $w \in W_{n,k}^+$, we have $\|e_w\|_2^2 = 9 \cdot 2^n$.

Define a sequence $(\phi_n)_{n=4}^{\infty}$ of linear functionals on $L(X_p)$ by $\phi_n(T)$ = mean trace of T over $\{e_w : w \in \bigcup_{k=1}^{2^n} W_{n,k}^+\} = 2^{-2n+1} \Sigma \{e_w^*(Te_w) : w \in \bigcup_{k=1}^{2^n} W_{n,k}^+\}$. We will define ϕ on $L(X_p)$ by $\phi(T) = \lim_{n \to \infty} \phi_n(T)$ and find a compact set K in X_p such that ϕ, K satisfy the conditions of Problem 4.1. Now $(\|e_w\|)$ is certainly not bounded in $l_p(\Gamma)$ for $p < \infty$, but one easily checks that $\|e_w\|_p \cdot \|e_w^*\|_q$ is bounded $((1/p) + (1/q) = 1)$, so it follows from Problem 4.2 that $\phi_n(T) \to 0$ for T compact. Of course, the boundedness $\|e_w\|_p \cdot \|e_w^*\|_q$ also guarantees that $(\|\phi_n\|_p)$ is bounded, where $\|\phi_n\|_p$ is the norm of the linear functional ϕ_n in $L(X_p)^*$. These simple observations are summarized in the following problem.

PROBLEM 4.11: *We have $\|e_w\|_\infty = 1$ and $\|e_w\|_p^p = 9 \cdot 2^n$ for $w \in W_{n,j}^+$ and $1 \le p < \infty$. Consequently, $\|e_w\|_p \|e_w^*\|_q = 1$ (for $1 \le p \le \infty$, $(1/p) + (1/q) = 1$), each ϕ_n is a norm one linear functional on $L(X_p)$ ($2 \le p \le \infty$), and $\lim_{n \to \infty} \phi_n(T) = 0$ for each compact operator T on X_p.*

The compact set K in X_p which satisfies the condition in Problem 4.1 will actually satisfy the hypothesis of:

PROBLEM 4.12: *Suppose that K is a compact set in X_p such that for $n = 4, 5, \ldots$, and $T \in L(X_p)$, we have $|\phi_n(T) - \phi_{n-1}(T)| \le \alpha_n \sup\{\|Tx\|_p : x \in K\}$, where $\phi_3 = 0$ and $\sum_{n=4}^{\infty} \alpha_n < \infty$. Then $\lim_{n \to \infty} \phi_n(T) = \phi(T)$ exists for each $T \in L(X_p)$, and ϕ, K satisfy the conditions in Problem 4.1.*

We come now to the problem of how to construct K. For motivation, again consider an operator T on the space $l_\infty(G)$, where $G = \{-1, 1\}^n$, let $W^+ \cup W^-$ be a decomposition of the characters W on G as in Problem 4.6, and let

$$y = \sum_{w \in W^+} w - \sum_{w \in W^-} w.$$

In the discussion after Problem 4.6, it was observed that if T is an operator on $l_\infty(G)$, and if $\phi(T) =$ mean trace of T over W^+ and $\psi(T) =$ mean trace of T over W^-, then

$$|\phi(T) - \psi(T)| \leq 2^{-[n/2]+2} \|T\|.$$

The point of the gluing procedure used to construct X_p is that $\phi_n(\cdot)$ simultaneously acts like the mean trace over $\bigcup_k W_{n,k}^+$ and the mean trace over $\bigcup_k W_{n+1,k}^-$. That is, $(\phi_{n+1} - \phi_n)(T)$ looks like the mean trace of T over $\bigcup_k W_{n+1,k}^+$ minus the mean trace of T over $\bigcup_k W_{n+1,k}^-$, which should be small relative to $\|T\|$; more precisely, that

$$|\phi_{n+1}(T) - \phi_n(T)| \leq \sup\{\|Tx\| : \|x\| \leq \delta,$$
$$x \in \mathrm{span}\{e_w : w \in \bigcup_k W_{n,k}^+\}\},$$

for a small value of δ.

In order to make this precise, it is helpful to write each ϕ_n in two different ways. For $n \geq 5$ and $1 \leq k \leq 2^n$, define $\phi_{n,k}, \psi_{n,k}$ on $L(X_p)$ by:

$$\phi_{n,k}(T) = \text{mean trace of } T \text{ over } \{e_w : w \in W_{n,k}^+\}$$
$$= 2^{-n+1} \sum_{w \in W_{n,k}^+} e_w^* T e_w;$$

$$\psi_{n,k}(T) = \text{mean trace of } T \text{ over } \{e_w : w \in hW_{n,k}^-\}$$
$$= 2^{-n+1} \sum_{w \in W_{n,k}^-} e_{h(w)}^* T e_{h(w)}.$$

PROBLEM 4.13: *We have*

$$2^{-n}\sum_{k=1}^{2^n}\phi_{n,k}=\phi_n=2^{-n-1}\sum_{k=1}^{2^{n+1}}\psi_{n+1,k},$$

so that

$$\phi_n-\phi_{n-1}=2^{-n}\sum_{k=1}^{2^n}(\phi_{n,k}-\psi_{n,k}).$$

It remains to estimate $\phi_{n,k}(T)-\psi_{n,k}(T)$ for fixed $n \geq 5$ and $1 \leq k \leq 2^n$. It is not possible to reduce to the case of diagonal operators T by averaging over operators U_g as in the motivational Problem 4.8, but we can accomplish about the same effect by evaluating T at points of X_p related to $U_g y$ for $g \in G_{n,k}$, where y is as in Problem 4.6. To this end, define $J=J_{n,k}$ for $w \in W_{n,k}$ by

$$Jw = e_w \quad \text{if } w \in W_{n,k}^+,$$
$$= e_{h(w)} \quad \text{if } w \in W_{n,k}^-,$$

and extend J linearly to an operator from span $W_{n,k}$ into X_p. Let $y = y_{n,k} = \sum_{w \in W_{n,k}^+} w - \sum_{w \in W_{n,k}^-} w$ be as in Problem 4.8, and set $A_{n,k} = \{2^{1-n} J U_g y : g \in G_{n,k}\}$.

PROBLEM 4.14: *For $T \in L(X_p)$, we have*

$$\phi_{n,k}(T)-\psi_{n,k}(T)=2^{1-n}|G_{n,k}|^{-1}\sum_{g \in G_{n,k}}(TJU_g y)(g)$$

and hence $|\phi_{n,k}(T)-\psi_{n,k}(T)| \leq \max\{\|Tx\|_p : x \in A_{n,k}\}$.

Hint: Approximate TJw by an element in span(e_w) to verify that $\langle TJw, w \rangle = e_w^*(TJw)\langle w, w \rangle$ if $w \in W_{n,k}^+$, and that $\langle TJw, w \rangle = e_{h(w)}^*(TJw)\langle w, w \rangle$ if $w \in W_{n,k}^-$. Then use Problem 4.7. □

The last important step is to estimate $\max\{\|x\|_p : x \in A_{n,k}\}$. Here, at last, the combinatorial conditions in Problem 4.9 are used.

PROBLEM 4.15: *We have* $\|JU_g y\|_2^2 = 27 \cdot 2^{2n-2}$ *for each* $g \in G_{n,k}$.

Hint: $U_g w = w(g) w$ by Problem 4.7, and $\|e_w\|_2$ is estimated in Problem 4.10.

PROBLEM 4.16: $\|JU_g y\|_\infty \leq 2 \cdot 2^{n/2}$ *for each* $g \in G_{n,k}$.

Hint: We have

$$JU_g y = J(\Sigma\{w(g)w : w \in W_{n,k}^+\} - \Sigma\{w(g)w : w \in W_{n,k}^-\})$$

$$= \Sigma\{w(g)w : w \in W_{n,k}^+\} + \Sigma\{\pm z : z \in h^{-1}(W_{n,k}^+)\}$$

$$\quad - \Sigma\{\pm w : w \in h(W_{n,k}^-)\} - \Sigma\{w(g)w : w \in W_{n,k}^-\}$$

$$\quad - \Sigma\{\pm z : z \in h^{-1}(h(W_{n,k}^-)) \sim W_{n,k}^-\}$$

$$\equiv z_1 + z_2 - z_3 - z_4 - z_5.$$

Note that $\|z_2\|_\infty = \|z_3\|_\infty = 1$. Problem 4.9 (ii) and (iv) imply that $\|z_5\|_\infty \leq 2$. Observe that $z_1 - z_4 = U_g y$, so that by Problem 4.6, we have $\|z_1 - z_4\|_\infty \leq 2 \cdot 2^{n/2}$.

PROBLEM 4.17: *For* $2 < p < \infty$, *we have*

$$\|JU_g y\|_p \leq \|JU_g y\|_2^{2/p} \|JU_g y\|_\infty^{1-2/p} \leq 3 \cdot 2^{n(1/2 + 1/p)}.$$

Hence, $\max\{\|x\|_p : x \in A_{n,k}\} \leq 6 \cdot 2^{n(1/p - 1/2)}$.

Finally, it is time to define the compact set K and the sequence (α_n) which satisfy the conditions in Problem 4.12.

Fix $2 < p < \infty$ and choose $\alpha_n > 0$ so that $\sum_{n=5}^\infty \alpha_n < \infty$ and $\lim_{n \to \infty} \alpha_n^{-1} 6 \cdot 2^{n(1/p - 1/2)} = 0$. Let $K = \{0\} \cup \bigcup_{n=5}^\infty \bigcup_{k=1}^{2^n} \alpha_n^{-1} A_{n,k}$.

PROBLEM 4.18: *The set* K *is a compact subset of* X_p. *For each* $T \in L(X_p)$, *and* $n = 5, 6, \ldots$, *we have*

$$|\phi_n(T) - \phi_{n-1}(T)| \leq \alpha_n \sup\{\|Tx\|_p : x \in K\}.$$

Added in proof: A. Szankowski has proved several results related to the topic treated in Section 4:

(1) For each $1 \leq p < 2$, l_p has a subspace which fails the approximation property.

(2) For each $2 < p < \infty$, there is a Banach lattice which embeds into $L_p(0,1)$ and which fails the approximation property [23].

(3) The space of bounded linear operators on l_2 fails the approximation property.

REFERENCES

1. A. M. Davie, "The approximation problem for Banach spaces," *Bull. London Math. Soc.*, **5** (1973), 261–266.
2. P. Enflo, "A counterexample to the approximation problem," *Acta. Math.*, **13** (1973), 309–317.
3. T. Figiel, "Further counterexamples to the approximation problem" (preprint).
4. T. Figiel and W. B. Johnson, "The approximation property does not imply the bounded approximation property," *Proc. Amer. Math. Soc.*, **41** (1973), 197–200.
5. T. Figiel and A. Pełczyński, "On Enflo's method of construction of Banach spaces without the approximation property," *Uspehi Mat. Nauk.*, **28** (1973), 95–108.
6. A. Grothendieck, "Produits tensoriels topologiques et espaces nucléaires," *Mem. Amer. Math. Soc.*, **16** (1955).
7. W. B. Johnson, "A complementably universal conjugate Banach space and its relation to the approximation problem," *Israel J. Math.*, **13** (1972), 301–310.
8. W. B. Johnson and A. Szankowski, "Complementably universal Banach spaces," *Studia Math.*, **58** (1976), 91–97.
9. M. I. Kadec, "On complementably universal Banach spaces," *Studia Math.*, **40** (1971), 85–89.
10. J. L. Kelley, *General Topology*, D. Van Nostrand, Princeton, 1955.
11. D. R. Lewis and C. Stegall, "Banach spaces whose duals are isomorphic to $l_1(\Gamma)$," *J. Functional Analysis*.
12. J. Lindenstrauss and L. Tzafriri, "On the complemented subspaces problem," *Israel J. Math.*, **9** (1971), 263–269.
13. H. Milne, "Banach space properties of uniform algebras," *Bull. London Math. Soc.*, **4** (1972), 323–326.
14. B. S. Mitjagin and N. M. Zobin, "Examples of nuclear Fréchet spaces without bases," *Funkcional. Anal. i Priložen.*, **8** (1974), 35–47 (Russian).
15. A. Pełczyński, "Projections in certain Banach spaces," *Studia Math.*, **19** (1960), 209–228.
16. _____, "Universal bases," *Studia Math.*, **32** (1969), 247–268.
17. _____, "Any separable Banach space with the bounded approximation

property is a complemented subspace of a Banach space with a basis," *Studia Math.*, **40** (1971), 239–242.
18. A. Pełczyński and Z. Semadeni, "Spaces of continuous functions III, Spaces $C(\Omega)$ for Ω without perfect subsets," *Studia Math.*, **18** (1959), 211–222.
19. A. Pełczyński and P. Wojtaszczyk, "Banach spaces with finite dimensional expansions of identity and universal bases of finite dimensional subspaces," *Studia Math.*, **40** (1971), 91–108.
20. H. P. Rosenthal, "On factors of $C[0,1]$ with non-separable dual," *Israel J. Math.*, **13** (1972), 361–378.
21. G. Schechtman, "On Pełczyński's paper 'Universal bases,'" *Israel J. Math.*, **22** (1975), 181–184.
22. A. Wilansky, *Functional Analysis*, Blaisdell, Waltham, Mass., 1964.
23. A. Szankowski, A Banach lattice without the approximation property, *Israel J. Math.*, **24** (1976), 329–337.

INTEGRAL REPRESENTATIONS FOR ELEMENTS OF CONVEX SETS

R. R. Phelps

The importance of convex sets has been recognized since early in the development of functional analysis and they remain the focus of a great deal of current research. It would take an enormous book to give a comprehensive survey of convexity; what we will do here is examine a small but useful and illuminating part of the subject, centering about the Choquet representation theorem for metrizable compact convex sets. We will assume that the reader has studied some general topology and has completed a first course in integration theory and real analysis, including the Hahn-Banach and Riesz representation theorems.

1. CHOQUET'S THEOREM

Recall that a subset K of a real linear space is said to be *convex* provided

$$\alpha x + (1-\alpha)y \in K$$

The author would like to express his thanks to Erik Alfsen, Niel Bogue, Richard Bourgin, Alexander Pełczyński, and David Pengelley, who read and criticized the first draft of his manuscript. He would also like to thank Boston Marathoner and artist, Mary Miller, who did the drawings.

whenever $x, y \in K$ and $0 \leq \alpha \leq 1$. More generally, an induction argument shows that K is convex if and only if *convex combinations* of elements of K are again in K, that is, if and only if

$$\sum_{k=1}^{n} \alpha_k x_k \in K$$

whenever $x_k \in K$ and $\alpha_k \geq 0$ for $k = 1, 2, \ldots, n$ and $\Sigma \alpha_k = 1$. Obvious examples of convex sets are (the regions interior to) triangles, squares, and circles in the plane $\mathbf{R}^2$. The *vertices* of a triangle (or of a square) and the boundary points of a circle are somewhat special, having the property that none of them is a relative interior point of a nontrivial segment lying in the set. Any point $x \in K$ with this property is called an *extreme point* of K, and the set of all such points is denoted by ext K. More precisely, if K is a convex set in a linear space, a point x of K is in ext K if and only if the only possible representation of x of the form

$$x = \tfrac{1}{2}(y + z), \quad y, z \in K$$

is the trivial one where $y = z = x$.

Any point x in a triangle can be expressed as a convex combination $x = \alpha_1 v_1 + \alpha_2 v_2 + \alpha_3 v_3$ of the vertices, the coefficients α_k being the standard "barycentric coordinates" of x. It follows easily from this that any point of a convex polygon can also be expressed as a convex combination of three (or fewer) vertices. For a circle, a given point can be similarly represented, using at most two extreme points. These observations are all special cases of Minkowski's theorem:

If K is a compact convex subset of a finite dimensional vector space, then each point of K is a convex combination of extreme points of K. More precisely (Carathéodory), if K is contained in an n-dimensional space, then each point of K can be expressed as a convex combination of $n + 1$ or fewer extreme points.

In specific examples, the extreme points of K often admit a particularly simple description; Minkowski's theorem then gives us a straightforward way of describing an arbitrary point of K. (We will sketch the proof of this theorem and look at one such example in Section 4.) Now, there are many compact convex sets which arise naturally in classical analysis, probability theory, potential theory, mathematical physics, and functional analysis, but these sets are seldom finite dimensional so the Minkowski theorem is no longer applicable. One substitute is the Krein-Milman [Kreĭn-Mil'man] theorem, which states that (under appropriate hypotheses) every point in such a set can be approximated as closely as desired by convex combinations of extreme points. This result has considerable utility (in part because it assures us that ext K is nonempty), but it does not give a representation theorem of the kind we seek. On the other hand, the Choquet theorem *does* give an integral representation and it, too, implies that ext K is nonempty. In order to motivate the Choquet theorem, we first reformulate Minkowski's theorem. Before doing this, we introduce some notation which will be used throughout.

Let X be a nonempty set. (In most of what follows, X will be a compact metric space.) For each $x \in X$ let ϵ_x denote the *point mass at* x; that is, ϵ_x is the measure defined on all subsets of X which satisfies, for any $A \subseteq X$,

$$\epsilon_x(A) = 1 \quad \text{if } x \in A,$$
$$= 0 \quad \text{if } x \notin A.$$

Since $\epsilon_x(X) = 1$ and $\epsilon_x(A) \geq 0$ for every $A \subseteq X$, the measure ϵ_x is a special case of a *probability measure*. An equivalent definition arises by describing what happens when we integrate a real-valued function f with respect to ϵ_x; namely,

$$\int f d\epsilon_x = f(x) \quad \text{for each such function } f.$$

We define convex combinations of point masses in a straightforward way: If $x_1, x_2, \ldots, x_p \in X$ and $\lambda_1, \lambda_2, \ldots, \lambda_p$ are positive numbers with $\Sigma \lambda_k = 1$, define $\mu = \Sigma \lambda_k \epsilon_{x_k}$ by setting

$$\int f d\mu = \sum \lambda_k f(x_k),$$

whenever f is a real-valued function on X. The measure μ is also a probability measure, and is called a *discrete* probability measure. As a set function it is characterized by

$$\mu(A) = \sum \{\lambda_k : x_k \in A\} \quad \text{for each } A \subseteq X.$$

In particular, its value at the singleton set $\{x_k\}$ is λ_k.

Returning to Minkowski's theorem, suppose that E denotes a finite dimensional Euclidean space and that K is a nonempty compact convex subset of E. If $x \in K$, then we can write $x = \sum_1^n \alpha_k x_k$, with each $\alpha_k > 0$, $\sum_1^n \alpha_k = 1$ and $\{x_k\}_{k=1}^n \subseteq \text{ext} K$, as described above. Let μ be the corresponding convex combination of the point masses ϵ_{x_k} defined by

$$\mu = \sum_1^n \alpha_k \epsilon_{x_k};$$

then μ is a probability measure on K and if f is any real-valued *linear* function on E (that is, if f is a member of the dual space E^*), then

$$\int f d\mu \equiv \mu(f) = \sum_1^n \alpha_k f(x_k) = f\left(\sum_1^n \alpha_k x_k\right) = f(x).$$

The assertion that $\mu(f) = f(x)$ for each $f \in E^*$, is described by saying that *the probability measure μ represents x*. We also say (using equivalent terminology) that *x is the resultant of μ*, or is the *barycenter* of μ. Note also that μ *is supported by* $\text{ext} K$, that is,

$$\mu(\text{ext} K) = \sum_1^n \alpha_k \epsilon_{x_k}(\text{ext} K) = \sum_1^n \alpha_k = 1.$$

Thus, it follows from Minkowski's theorem that *to each $x \in K$ there corresponds a discrete probability measure μ on $\text{ext} K$ which represents x*. It is readily seen that this is, in fact, a reformulation of Minkowski's theorem. Indeed, if $x \in K$ and if x is represented by a discrete probability measure μ on $\text{ext} K$, then μ has the above form for some finite subset $\{x_k\} \subseteq K$ and positive numbers α_k; moreover, $\mu(\text{ext} K) = 1$. Since $\mu(\{x_k\}) = \lambda_k > 0$ for each k, it follows that $x_k \in \text{ext } K$. Finally, the convex combination $\Sigma \alpha_k x_k$ coincides with x; if not, we could find f in E^* such that $f(x) \neq f(\Sigma \alpha_k x_k)$, contradicting the fact that the right side is $\Sigma \alpha_k f(x_k) = \mu(f) = f(x)$.

The above reformulation of Minkowski's theorem suggests a possible generalization to infinite dimensional K: Show that to each $x \in K$ there corresponds at least one probability measure μ on ext K (not necessarily discrete) which represents x. This is exactly what the Choquet theorem does. We must be more precise, of course, in saying what kinds of sets and measures are involved, and what is meant by the word "represents."

The most immediate infinite dimensional generalization of Euclidean space is the separable Hilbert space l_2 of all square summable real sequences. Since any two separable Hilbert spaces are equivalent, we will use the abstract version of l_2, letting H denote, henceforth, a separable real Hilbert space provided with the inner product (,) and norm given by $\|x\|^2 = (x,x)$. Unless specified otherwise, K will always denote a nonempty compact convex subset of H. An *affine map* T from a convex set A into a linear space E is a map which preserves convex combinations; equivalently,

$$T[\alpha x + (1-\alpha)y] = \alpha T(x) + (1-\alpha)T(y)$$

whenever $x, y \in A$ and $0 \leq \alpha \leq 1$. [For those readers familiar with locally convex Hausdorff linear spaces, we hasten to point out that any compact convex metrizable subset of such a space admits an affine homeomorphism T onto a compact convex subset $K \subseteq H$; since T^{-1} is also affine, T itself preserves extreme points and we will have lost no generality in restricting ourselves to such sets K. The existence of T can be proved using the ideas contained in the proof of Proposition 2.1 below.]

Since K is a subset of H, it is, in particular, a compact metric space. In order to discuss probability measures on K, it is simplest to discuss measures and continuous functions on an arbitrary compact Hausdorff space X. We denote by $C(X)$ the Banach space of all real valued continuous functions on X, with the usual supremum norm:

$$\|f\| = \sup\{|f(x)| : x \in X\}, \quad f \in C(X).$$

The *Borel subsets* of X are defined to be the smallest σ-algebra of subsets of X which contains all the closed subsets. By a *measure* on X we will mean a nonnegative finite measure μ on the Borel sets that is *regular* in the sense that, for each Borel set $B \subseteq X$, we have

$$\mu(B) = \sup\{\mu(A) : A \text{ compact}, A \subseteq B\}.$$

A *probability* measure μ is one that satisfies $\mu(X) = 1$. The *Riesz representation theorem* states that if L is a linear functional on $C(X)$ that is *positive* in the sense that $L(f) \geqslant 0$ whenever $f \in C(X)$ and $f \geqslant 0$, then there is a unique measure μ on X such that

$$L(f) = \int_X f \, d\mu, \quad f \in C(X).$$

The measure μ will be a probability measure if and only if L takes the value 1 at the function that is identically 1. (We will denote any constant function by the number which is its constant value; thus μ is a probability measure if and only if $L(1) = 1$.)

Here is a basic and useful fact about $C(X)$ which is omitted from most textbooks; it can be proved by an elegant application of the Stone-Weierstrass theorem [6, p. 134]: *If X is a compact metric space, then $C(X)$ is separable.*

Returning to our compact convex metric space K, define $A(K) \subseteq C(K)$ to be the subspace of all *continuous real-valued affine functions* on K. It is easily checked that $A(K)$ is a uniformly closed (=norm closed) linear subspace of $C(K)$ which contains the constant functions. Moreover, $A(K)$ *separates points of K*; indeed, if $x, y \in K$ and $x \neq y$, then $0 < \|x-y\|^2 = (x-y, x-y) = (x,z) - (y,z)$, where we have set $z = x - y$. Since the function

$$u \mapsto (u, z)$$

is continuous and linear on H, its restriction to K is affine and continuous, and takes different values at x and y.

We can now give a precise formulation of the Choquet theorem.

THEOREM 1.1 (Choquet): *Let K be a compact convex subset of the separable real Hilbert space H. Then for any $x \in K$ there exists a*

probability measure μ on K such that
  (i) $\int_K f d\mu = f(x)$ *for each f in $A(K)$, and*
  (ii) $\mu(\text{ext} K) = 1$.

This obviously resembles closely the reformulated version of Minkowski's theorem. One slight difference is the use of functions in $A(K)$ in place of the smaller class of those affine functions which are restrictions to K of continuous linear functionals on the entire space. This is less general than it appears, since such restrictions (together with the constant functions) constitute a dense subspace of $A(K)$. (This fact, proved in [12, Prop. 4.5], is not needed here.) We should say a word about part (ii); it obviously does not make sense unless $\text{ext} K$ is a measurable set. As the next result shows, however, $\text{ext} K$ is actually a Borel set.

PROPOSITION 1.2: *For any metrizable compact convex set K, the set $\text{ext} K$ is a G_δ subset of K.*

Proof: Let d denote any continuous metric on K and, for each $n \geq 1$, define F_n to be the set of all points $x \in K$ for which there exist points y, z in K satisfying

$$x = \tfrac{1}{2}(y+z) \quad \text{and} \quad d(y,z) \geq 1/n.$$

The compactness of K and the continuity of d readily yield the fact that each set F_n is closed, and it follows from the definition of an extreme point that $K \setminus \text{ext} K = \bigcup F_n$, so that $\text{ext} K$ is a G_δ set.

It is trivial to find a probability measure μ on K which satisfies (i); simply let $\mu = \epsilon_x$. Part (ii) fails, however, unless we started with x in $\text{ext} K$. If $x \notin \text{ext} K$, then it can be expressed as $\tfrac{1}{2}(x_1 + x_2)$, where $x_1 \neq x_2$. The corresponding discrete measure $\mu = \tfrac{1}{2}(\epsilon_{x_1} + \epsilon_{x_2})$ satisfies (i) and, if x_1, x_2 are extreme points, will satisfy (ii). If $x_1 \notin \text{ext} K$, say, we can decompose it in the same way, getting x as a convex combination of three points which are, in some sense, "closer" to being extreme than is x. Presumably, by continuing this decomposition procedure we could, "in the limit," get a representing measure on $\text{ext} K$. This kind of argument can be made to work, but a much simpler procedure can be followed once we have introduced the notion of the upper envelope of a convex function.

DEFINITION: A real valued function f on a convex set K is said to be *convex* if

$$f[\alpha x+(1-\alpha)y] \leqslant \alpha f(x)+(1-\alpha)f(y)$$

whenever $x, y \in K$ and $0 \leqslant \alpha \leqslant 1$. We say that f is *concave* if the inequality goes the other way, that is, if the function $-f$ is convex. We say that f is *strictly convex* if

$$f[\alpha x+(1-\alpha)y] < \alpha f(x)+(1-\alpha)f(y)$$

whenever $x \neq y$ and $0 < \alpha < 1$.

The restriction to K of the square of the norm in the Hilbert space H is an important example of a strictly convex continuous function. Indeed, if $x \neq y$, then

$$0 < \|x-y\|^2 = (x-y, x-y) = \|x\|^2 - 2(x,y) + \|y\|^2.$$

Using this, one can verify that for $0 < \alpha < 1$, we have

$$\|\alpha x+(1-\alpha)y\|^2 = \alpha^2\|x\|^2 + 2\alpha(1-\alpha)(x,y) + (1-\alpha)^2\|y\|^2$$
$$< \alpha\|x\|^2 + (1-\alpha)\|y\|^2.$$

DEFINITION: Let f be a bounded real valued function on the set K. Define the *upper envelope* $\bar{f}$ of f by

$$\bar{f}(x) = \inf\{h(x) : h \in A(K), h \geqslant f\}, \qquad x \in K.$$

It is immediate from the continuity of the functions in $A(K)$ and the definition of $\bar{f}$ that the latter is always *upper semicontinuous*; that is, for each real number λ, the set

$$\{x \in K : \bar{f}(x) < \lambda\}$$

is open. Since open sets are Borel sets, this also shows that the function $\bar{f}$ is Borel measurable. It also follows easily from the definition of $\bar{f}$ and from the fact that the functions in $A(K)$ are affine that *the function $\bar{f}$ is always concave*. Since $A(K)$ contains the constant functions, any upper bound for f is an upper bound

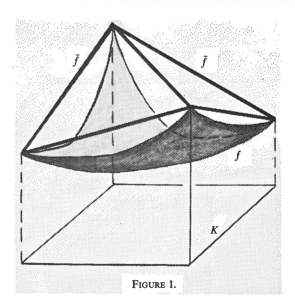

FIGURE 1.

for $\bar{f}$, while $f \leq \bar{f}$ shows that $\bar{f}$ is bounded below. Thus, $\bar{f}$ is always a bounded Borel measurable function, hence integrable with respect to any finite Borel measure.

In order to visualize the function $\bar{f}$ a little better, it helps to consider what it must be when f is a convex function defined on a compact convex subset K of the plane. First, if f is actually affine, then $\bar{f} = f$ and its graph lies in a plane, no matter what the set K is. If K is a triangle, then a little thought shows that for any convex f, the graph of $\bar{f}$ will lie in the plane determined by the three points of the graph of f corresponding to the vertices of K. If K is a square, then the graph of $\bar{f}$ *can* lie in a plane, but in general will consist of two triangular pieces, again determined by the values of f at the vertices of K, as in Figure 1. In each of these examples, the functions f and $\bar{f}$ take the same values at the extreme points of K. This is *always* the case, but we do not need this fact. We *do* need its converse, however, which is valid for strictly convex functions.

PROPOSITION 1.3: *If f is a strictly convex continuous function on K and if $f(x) = \bar{f}(x)$, then $x \in \text{ext}\, K$.*

Proof: If $x \in K$ is *not* an extreme point, then we can write $x = \frac{1}{2}(y+z)$, where y and z are distinct elements of K. Using the strict convexity of f and the concavity of $\bar{f}$, we have

$$f(x) < \tfrac{1}{2}[f(y)+f(z)] \leq \tfrac{1}{2}[\bar{f}(y)+\bar{f}(z)] \leq \bar{f}(x),$$

which completes the proof.

This shows that if f is strictly convex and continuous, then any probability measure μ on K which satisfies $\mu(f) = \mu(\bar{f})$ is necessarily supported by ext K; indeed, since $\bar{f} - f \geq 0$, the set $\{x \in K: \bar{f}(x) > f(x)\}$ must have μ measure zero. Consequently,

$$1 = \mu(\{x \in K: f(x) = \bar{f}(x)\}) \leq \mu(\text{ext}\,K) \leq 1.$$

To prove Choquet's theorem, then, we need only produce a probability measure μ with resultant x which satisfies $\mu(f) = \mu(\bar{f})$ for some strictly convex continuous function f on K. Before doing this, let us reexamine the intuitively appealing decomposition procedure described earlier and see why it should lead to a representing measure μ with the above property. If a point $x \in K$ can be decomposed as a nontrivial convex combination of other points of K, say $x = \sum_1^n \alpha_i x_i$, then the strict convexity of f, the concavity of $\bar{f}$, and the fact that $f \leq \bar{f}$, together imply (exactly as in the proof of the above proposition) that

$$f(x) < \sum_1^n \alpha_i f(x_i) \leq \sum_1^n \alpha_i \bar{f}(x_i) \leq \bar{f}(x).$$

In terms of the measure ϵ_x and the discrete measure $\mu = \sum_1^n \alpha_i \epsilon_{x_i}$, we have

$$\epsilon_x(f) < \mu(f) \leq \mu(\bar{f}) \leq \bar{f}(x).$$

Now, if we additionally decompose one of the points x_i and make the appropriate substitution so as to express x as the corresponding convex combination of this larger set of points, then the corresponding discrete measure λ will still have resultant x and will necessarily satisfy $\mu(f) < \lambda(f)$ and $\lambda(\bar{f}) \leq \mu(\bar{f})$. (Simply apply the above kind of inequality to x_i instead of to x.) For instance, if $x = \frac{1}{2}(x_1 + x_2)$ and if we can write $x_1 = \frac{1}{2}(x_3 + x_4)$, then we have $\mu = \frac{1}{2}(\epsilon_{x_1} + \epsilon_{x_2})$ and

$$\lambda = \frac{1}{2}\left[\frac{1}{2}(\epsilon_{x_3} + \epsilon_{x_4})\right] + \frac{1}{2}\epsilon_{x_2},$$

so λ represents x and

$$\lambda(f) = \tfrac{1}{4}f(x_3) + \tfrac{1}{4}f(x_4) + \tfrac{1}{2}f(x_2) > \tfrac{1}{2}[f(x_1) + f(x_2)] = \mu(f).$$

Thus, the decomposition procedure yields a sequence of measures $\{\mu_n\}$ such that $\{\mu_n(f)\}$ is increasing and $\{\mu_n(\bar{f})\}$ is nonincreasing while $\mu_n(f) \leq \mu_n(\bar{f})$ for each n. Both sequences of numbers necessarily converge (but possibly not to the same value). By modifying the process above, however, one can get a limit measure μ such that $\mu(f) = \mu(\bar{f})$. This requires more machinery than we have at present; instead, a rather simple application of the Hahn-Banach theorem allows us to bypass the decomposition procedure completely.

THEOREM (Choquet): *For each $x \in K$ there exists a probability measure μ with resultant x such that $\mu(\text{ext} K) = 1$.*

Proof: Let f be a strictly convex continuous function on K and let B be the subspace of $C(K)$ consisting of all functions of the form $h + rf$, $h \in A(K)$, r real. Given $x \in K$, define

$$\alpha = \inf\{h(x) + \|h - f\| : h \in A(K)\}.$$

We can show that this is finite (and prove a useful inequality) by showing that

$$g(x) - \|g - f\| \leq \alpha$$

whenever $g \in A(K)$. Indeed, if $h \in A(K)$, then $g(x) - h(x) \leq \|g - h\| \leq \|g - f\| + \|h - f\|$, so

$$g(x) - \|g - f\| \leq h(x) + \|h - f\|;$$

now take the infimum of the right-hand side. Next, define L on B

by

$$L(h+rf) = h(x) + r\alpha \quad (h \in A(K), r \text{ real}).$$

This is an extension to B of the evaluation functional L_x on $A(K)$, so $L(1)=1$. We also have $\|L\|=1$. To show this we need only show that $\|L\| \leq 1$; that is,

$$h(x) + r\alpha \equiv L(h+rf) \leq \|h+rf\| \quad (h \in A(K), r \text{ real}).$$

If $r>0$, this is equivalent (writing $g = -r^{-1}h$) to

$$-g(x) + \alpha \leq \|g-f\|$$

which is immediate from the definition of α. The case $r=0$ is obvious, while if $r<0$, the desired inequality is equivalent (again writing $g = -r^{-1}h$) to

$$g(x) - \alpha \leq \|g-f\|$$

which is in turn equivalent to the useful inequality proved above.

By the Hahn-Banach theorem, we can extend L to a functional L' of norm 1 on all of $C(K)$. Since $\|L'\|=1$ and $L'(1)=1$, we have $L'(g) \geq 0$ whenever $g \in C(K)$ and $g \geq 0$. [Indeed, if $g \geq 0$, then $0 \leq \|g\| - g \leq \|g\|$, which shows that

$$\|(\|g\|-g)\| \leq \|g\|.$$

Consequently

$$\|g\| - L'(g) = L'(\|g\|-g) \leq \|(\|g\|-g)\| \leq \|g\|,$$

and therefore $L'(g) \geq 0$.] By the Riesz representation theorem there exists a probability measure μ on K which represents L'; in particular, for $h \in A(K)$, we have

$$\int_K h \, d\mu = L'(h) = L(h) = h(x).$$

Also, if $h \in A(K)$, then (as shown above) $g = h + \|h-f\| \geq f$ and $g \in A(K)$, so

$$\mu(g) = h(x) + \|h-f\| \geq \inf\{\mu(h'): h' \in A(K), h' \geq f\}.$$

If $h' \in A(K)$ and $h' \geq f$, then $h' \geq \bar{f}$ so $\mu(h') \geq \mu(\bar{f})$ and therefore the right side of the above inequality dominates $\mu(\bar{f})$. The infimum of the left side over all h in $A(K)$ is precisely $\alpha = L(f) = \mu(f)$, so that $\mu(f) \geq \mu(\bar{f})$. Since $f \leq \bar{f}$, the reverse inequality is immediate and the proof is complete.

2. A COMMON SETTING FOR APPLICATIONS

Since most compact convex sets which arise in applications are subsets of the dual of a Banach space, in its weak* topology, we will review some of the basic facts about such spaces. If E is a real Banach space, we let E^* denote the space of all continuous linear functionals L on E, with norm $\|L\| = \sup\{|L(f)|: f \in E, \|f\| \leq 1\}$. An immediate corollary of the Hahn-Banach theorem shows that there are enough linear functionals in E^* to separate points of E. The weak* topology is easily described: A net (L_α) converges to L if and only if $\lim_\alpha L_\alpha(f) = L(f)$ for each $f \in E$. Equivalently, the weak* topology is the topology induced on E^* when it is considered as a subset of the Cartesian product space $\mathbf{R}^E$ with the product topology. (We let $\mathbf{R}$ denote the real numbers.) The usefulness of the weak* topology derives in large part from the following proposition.

PROPOSITION 2.1: *Let E be a normed linear space and suppose that K is a norm bounded weak* closed convex subset of E^*. Then K is weak* compact. If E is separable, then K (in its weak* topology) is affinely homeomorphic to a compact convex subset of the Hilbert space l_2.*

Proof: By hypothesis, there exists a constant $M > 0$ such that $\|L\| \leq M$ for each $L \in K$. For each $f \in E$ let I_f denote the closed real interval $[-M\|f\|, M\|f\|]$ and let X denote the Cartesian product of all the intervals $I_f, f \in E$. By the Tychonoff theorem, X is compact in the product topology. The canonical embedding $L \mapsto (L(f))_{f \in E}$ of E^* into $\mathbf{R}^E$ carries K into the subset X and (by definition of the weak* topology and the hypothesis that K is weak* closed) it follows that K is weak* compact.

Suppose, now, that E is separable, so that there exists a norm-dense sequence $\{f_n\} \subseteq E \setminus \{0\}$. For each $L \in K$, let

$$T(L) = (2^{-n} L(f_n)/\|f_n\|)_{n=1}^{\infty}.$$

The sequence $T(L)$ is an element of l_2, since $|L(f_n)| \leq M\|f_n\|$ for each n and therefore

$$\sum_1^{\infty} [2^{-n} L(f_n)/\|f_n\|]^2 \leq M^2 \sum_1^{\infty} 4^{-n} < \infty.$$

It is easy to check that T is affine and one-one (because $\{f_n\}$ is dense) and hence T^{-1} is affine. Using the boundedness of K and an $\epsilon/2$ argument, it is not hard to show that T is continuous from the weak* topology of K into the norm topology of l_2. Since K is weak* compact, we conclude that T is a homeomorphism between K and its image $T(K)$, which completes the proof.

The above proposition shows, in particular, that each weak* compact convex subset K of the dual of a separable Banach space is metrizable in the weak* topology; explicitly, if $L, L' \in K$, then we can define $d(L, L') = \|T(L) - T(L')\|$, where the norm on the right is the one in l_2. It is readily verified that the unit ball $U = \{L \in E^*: \|L\| \leq 1\}$ is weak* closed convex and, of course, norm bounded, so any weak* closed convex subset of U is weak* compact and—if E is separable—metrizable. A basic example is the set $P_1(X)$ of all probability measures on a compact metric space X; as mentioned in Section 1, the space $C(X)$ is separable, so $P_1(X)$ is a weak* metrizable subset of $C(X)^*$ and the Choquet theorem is applicable. As we shall see later, in this case the Choquet theorem is precisely the Riesz representation theorem!

There is a second useful property of a weak* compact convex subset K of the dual E^* of a normed (or Banach) space E: For each $f \in E$, the function from K to $\mathbf{R}$ defined by

$$L \mapsto L(f)$$

is affine and weak* continuous, hence is an element of $A(K)$. This fact is an obvious consequence of the definition of the weak* topology, but despite its simplicity, it is of great utility in applications of the Choquet theorem. As an easy illustration of this, we prove John Rainwater's characterization of weak convergence of bounded sequences in a Banach space. (Recall that a sequence $\{f_n\}$ in a Banach space E converges *weakly* to $f \in E$ provided $L(f_n) \to L(f)$ for each $L \in E^*$.)

THEOREM 2.2 (Rainwater): *Suppose that E is a separable Banach space, that $\{f_n\}$ is a bounded sequence in E and that $f_0 \in E$. Then $\{f_n\}$ converges weakly to f_0 if and only if $L(f_n) \to L(f_0)$ for each extreme point L of the unit ball U of E^*.*

Proof: The "only if" part is immediate from the definition of weak convergence. To prove the "if" portion, suppose $L_0 \in E^*$. We may assume (without loss of generality) that $\|L_0\| \leq 1$, so that $L_0 \in U$. From Proposition 2.1, the Choquet representation theorem is applicable to U, so there exists a probability measure μ on $\text{ext}\, U$ such that

$$h(L_0) = \int_{\text{ext}\, U} h(L)\, d\mu(L) \quad \text{for each } h \in A(U).$$

As noted above, for each $f \in E$ the function $L \mapsto L(f)$ ($L \in U$) is a member of $A(U)$, so we conclude that

$$L_0(f) = \int_{\text{ext}\, U} L(f)\, d\mu(L) \quad \text{for each } f \in E.$$

Since $\{f_n\}$ is bounded, we have $\|f_n\| \leq M$ for some $M > 0$, hence $|L(f_n)| \leq M$ for each $L \in U$. By hypothesis, $L(f_n) \to L(f_0)$ for each $L \in \text{ext}\, U$ and hence, by the Lebesgue dominated convergence theorem,

$$L_0(f_n) = \int_{\text{ext}\, U} L(f_n)\, d\mu(L) \to \int_{\text{ext}\, U} L(f_0)\, d\mu(L) = L_0(f_0).$$

Since this is true for each $L_0 \in E^*$, the proof is complete.

[It is not hard to generalize the above result to nonseparable Banach spaces E; one applies it to the separable subspace E_0

generated by $\{f_n\}$ and f_0, then uses the Hahn-Banach and Krein-Milman theorems to show that each extreme point of the (metrizable) unit ball of E_0^* is the restriction to E_0 of an extreme point of the unit ball of E^*.]

The foregoing application of the Choquet theorem was rather abstract and the utility of the latter did not depend on finding a characterization of the extreme points of the convex set U. This is in contrast to the situation which usually arises in applications, where a concrete description of the extreme points is all-important. The example mentioned above, of the set of all probability measures on a compact metric space X, provides an illustration of how the extreme points can have a fascinating and remarkable description. While this does not lead us to another application of Choquet's theorem, it is a basic result. Since it is easily proved without assuming that X is metrizable, we will drop that hypothesis.

PROPOSITION 2.3: *Suppose that X is a compact Hausdorff space and let P_1 denote the weak* compact convex set of all regular Borel probability measures on X. Equivalently, let P_1 be the set of all linear functionals L on $C(X)$ satisfying*

$$L \geq 0 \quad \text{and} \quad L(1) = 1.$$

Then the following assertions are equivalent for an element $L \in P_1$:
  (i) *L is an extreme point of P_1.*
  (ii) *L is multiplicative: $L(fg) = L(f)L(g)$ for all $f, g \in C(X)$.*
  (iii) *$L = \epsilon_x$ for some $x \in X$.*

Moreover, the map ϕ: $X \to P_1$ defined by $\phi(x) = \epsilon_x$ is a homeomorphism between X and $\operatorname{ext} P_1$ (the latter in the weak* topology).

Proof: (i) implies (ii). Suppose that $L \in \operatorname{ext} P_1$ and that $g \in C(X)$, with $0 \leq g \leq 1$. We will first show that $L(fg) = L(f)L(g)$ for each $f \in C(X)$. To see this, let M be the linear functional on $C(X)$ defined by

$$M(f) = L(fg) - L(f)L(g), \quad f \in C(X).$$

Clearly, $M(1) = 0$, so $(L \pm M)(1) = 1$. Moreover, if $f \geq 0$, then

$$(L + M)(f) = L(f)[1 - L(g)] + L(fg) \geq 0$$

since $1 - L(g) \geq 0$ and the other terms are nonnegative. Also,

$$(L - M)(f) = L[f(1 - g)] + L(f)L(g) \geq 0.$$

Thus, if $L_1 = L + M$ and $L_2 = L - M$, then $L_1, L_2 \in P_1$ and $L = \frac{1}{2}(L_1 + L_2)$. Since L is extreme, this implies that $L_1 = L = L_2$, which means that $M = 0$; that is, $L(fg) = L(f)L(g)$ whenever $0 \leq g \leq 1$. If g is any nonconstant function, let $a = \|g\|$ (so $g + a \geq 0$) and let $b = \|g + a\|^{-1}$, so $0 \leq b(g + a) \leq 1$. Applying what we have just shown to this latter function yields the desired conclusion, since (ii) holds trivially if g is constant.

(ii) implies (iii). Suppose that L is multiplicative but that for each $x \in X$, we have $L \neq \epsilon_x$; we will show that this leads to a contradiction. Since $L \neq \epsilon_x$ there exists $f \in C(X)$ such that $L(f) \neq f(x)$. Replacing f by $f - L(f)$ allows us to assume that $L(f) = 0 \neq f(x)$, while replacing f by f^2 [and using the fact that $L(f^2) = L(f)^2$] allows us to assume that $f \geq 0$, $L(f) = 0$ and $f(x) > 0$. Thus, for each $x \in X$ there exists $f_x \in C(X)$ such that $f_x \geq 0$, $L(f_x) = 0$ and $f_x(x) > 0$, and there exists an open neighborhood U_x of x in which $f_x > 0$. By compactness, we can cover X with finitely many such neighborhoods, which shows that there exist nonnegative functions $f_1, f_2, \ldots, f_n$ in $C(X)$ such that $L(f_i) = 0$ for each i while $f = \sum f_i$ is strictly positive. Thus, $f \geq m$ for some $m > 0$ and therefore $0 = \sum L(f_i) = L(f) \geq L(m) = m > 0$, the desired contradiction.

(iii) implies (i). Suppose that $x \in X$ and that $\epsilon_x = \frac{1}{2}(\mu_1 + \mu_2)$, where $\mu_1, \mu_2 \in P_1$. If $\mu_1 \neq \epsilon_x$, say, then $\mu_1(X \setminus \{x\}) > 0$ and hence (by regularity) there exists a compact subset $A \subseteq X \setminus \{x\}$ such that $\mu_1(A) > 0$. By Urysohn's lemma we can choose $f \in C(X)$ such that $f \geq 0$, $f(x) = 0$ and $f = 1$ on A. Then $\mu_1(f) > 0$ and $\mu_2(f) \geq 0$ so that

$$0 < \mu_1(f) + \mu_2(f) = 2\epsilon_x(f) = 0,$$

a contradiction which shows that $\mu_1 = \epsilon_x$ and hence $\epsilon_x \in \text{ext } P_1$.

Finally, to see that ϕ is a homeomorphism, one simply checks that the continuity of each $f \in C(X)$ implies that ϕ is continuous from X into P_1 (in the weak* topology); the fact that $C(X)$ separates points of X shows that ϕ is one-one and hence the compactness of X implies that ϕ is a homeomorphism from X onto $\phi(X) = \text{ext} P_1$. This completes the proof.

Suppose that X is compact metric, so that $K = P_1(X)$ is weak* compact, convex and metrizable, and hence Choquet's theorem is applicable. Thus, to each $L_0 \in K$ there corresponds a probability measure μ on $\text{ext} K$ such that for each $f \in C(X)$ we have

$$L_0(f) = \int_{\text{ext} K} L(f) \, d\mu(L).$$

Since $\phi: X \to \text{ext} K$ is a homeomorphism, the set function λ defined on the Borel subsets of X by $\lambda(B) = \mu(\phi(B))$ is a regular Borel probability measure on X. The above integral formula now becomes

$$L_0(f) = \int_X \phi(x)(f) \, d\mu(\phi(x)) = \int_X f(x) \, d\lambda(x)$$

which is the Riesz representation for L_0. One can reverse this argument: Starting with the Riesz theorem and carrying the resulting measure λ to a measure $\mu = \lambda \circ \phi^{-1}$ on $\text{ext} K$, one gets a Choquet-type representation. [While the Choquet theorem is more general than the Riesz theorem, it should be remembered that the latter played an important part in the proof of the former.]

3. AN APPLICATION: INVARIANT AND ERGODIC MEASURES

Ergodic theory, which has its foundations in statistical mechanics, can be described as the study of measure-preserving transformations. By a *measure-preserving* transformation we mean a map $T: \Omega \to \Omega$ of a measure space (Ω, Σ, μ) into itself with the property that $\mu(T^{-1}B) = \mu(B)$ whenever $B \in \Sigma$. (In order for this to make sense, it is assumed that T is *measurable*, in the sense that $T^{-1}B \in \Sigma$ whenever $B \in \Sigma$.) The transformation T is said to be *ergodic* if it does a thorough job of "stirring up" Ω. If T were one-one, we could define this more precisely by saying that the only measurable sets B which satisfy $TB = B$ are Ω and the empty set. Since one wants to consider more general maps T, we replace $TB = B$ by $T^{-1}B = B$ and, in order to ignore sets of measure zero, we replace the conclusion that $B = \varnothing$ or $B = \Omega$ by $\mu(B) = 0$ or $\mu(B) = \mu(\Omega)$, which we assume is finite. Finally, we can further ignore sets of measure zero if we replace $T^{-1}B = B$ by $\mu(T^{-1}B \Delta B) = 0$, where, for any two measurable sets B_1, B_2, the symmetric difference $B_1 \Delta B_2$ is defined to be the set $(B_1 \setminus B_2) \cup (B_2 \setminus B_1)$. Thus we arrive at the following definition.

DEFINITION: A measure-preserving measurable map T of the finite measure space (Ω, Σ, μ) is said to be *ergodic* provided that $\mu(B) = 0$ or $\mu(B) = \mu(\Omega)$ whenever $B \in \Sigma$ is such that $\mu(T^{-1}B \Delta B) = 0$.

One can approach the study of such transformations from a different viewpoint, by starting with a measurable transformation T (or, more generally, a family of measurable transformations) of a measurable space (Ω, Σ) and looking for probability measures on Σ with respect to which each of the transformations is invariant or ergodic. This approach leads us to the following definition.

DEFINITION: Let (Ω, Σ) be a measurable space and let $\mathcal{T}$ be a family of measurable transformations of Ω into itself. A probability measure μ on Σ is said to be *invariant with respect* to $\mathcal{T}$ (or simply *invariant*) if $\mu(T^{-1}B) = \mu(B)$ for each $B \in \Sigma$ and each $T \in \mathcal{T}$. An invariant probability measure μ on Σ is said to be *ergodic* if $\mu(B) = 0$ or $\mu(B) = 1$ whenever $B \in \Sigma$ is such that $\mu(T^{-1}B \Delta B) = 0$ for each T in $\mathcal{T}$.

For example, if X is a compact metric space and Σ is the σ-algebra of Borel subsets of X, then any continuous map $T: X \to X$ is measurable. If $\mathcal{T}$ consists of the identity mapping alone, then the set of invariant probability measures consists of the weak* compact convex set of *all* probability measures, and μ will be ergodic if and only if $\mu(B) = 0$ or $\mu(B) = 1$ for each Borel set B; that is, if and only if μ is a point mass ϵ_x for some $x \in X$. This shows (cf. Proposition 2.3) that in this special case the ergodic measures are the extreme points of the set of invariant measures. It is an interesting fact that this is true under very general circumstances. We state the general result and prove the easier direction. (Note that we do not assert that the set $\tilde{P}_1$, defined below, is nonempty.)

THEOREM: *Let $\mathcal{T}$ be any family of measurable transformations of the measurable space (Ω, Σ) into itself and let $\tilde{P}_1$ denote the set of all $\mathcal{T}$-invariant probability measures on (Ω, Σ). The extreme points of $\tilde{P}_1$ are precisely those members of $\tilde{P}_1$ which are ergodic.*

Partial proof: One part of the proof is fairly easy: Suppose that $\mu \in \tilde{P}_1$ and that $A \in \Sigma$ is such that $0 < \mu(A) < 1$ and $\mu(T^{-1}A \Delta A) = 0$ for each $T \in \mathcal{T}$. For $B \in \Sigma$, define

$$\mu_1(B) = \mu(B \cap A)/\mu(A) \text{ and } \mu_2(B) = \mu(B \setminus A)/[1 - \mu(A)].$$

Both μ_1 and μ_2 are probability measures; moreover, they are invariant. For instance, if $T \in \mathcal{T}$, then the set theoretic identity

$$T^{-1}B \cap (T^{-1}A \Delta A) = (T^{-1}B \cap T^{-1}A) \Delta (T^{-1}B \cap A)$$

and the hypothesis on A show that the right side has μ measure zero. From this we conclude that

$$\mu(T^{-1}B \cap A) = \mu(T^{-1}B \cap T^{-1}A) = \mu(T^{-1}(B \cap A)) = \mu(B \cap A)$$

(by invariance of μ) and it follows immediately that μ_1 is invariant. Since we can write

$$\mu = \mu(A)\mu_1 + [1 - \mu(A)]\mu_2,$$

we see that μ_2 is also invariant; moreover, μ is a nontrivial convex combination of invariant measures, hence is not extreme.

The proof of the converse, while still elementary, requires more work; we refer the interested reader to [12, Section 10].

The question as to whether a given transformation admits an invariant measure is an interesting and sometimes difficult one.

However, if the family $\mathcal{T}$ is a commutative semigroup (under composition) of continuous maps of a compact Hausdorff space X into itself, then the Markov-Kakutani fixed-point theorem [4, p. 109] or [7, p. 456] yields the existence of at least one invariant probability measure. This is because each $T \in \mathcal{T}$ induces a weak* continuous affine map $\mu \mapsto T^*\mu$ of the probability measures on X, by the formula $(T^*\mu)(B) = \mu(T^{-1}B)$, B any Borel subset of X. It follows immediately that $T^*\mu = \mu$ if and only if μ is invariant, so the set of invariant measures coincides with the set of common fixed points for the maps T^*. These remarks can be combined with Choquet's theorem to yield the following result, which shows that each invariant probability measure is an integral average of ergodic measures.

THEOREM: *Let X be a compact metric space and let $\mathcal{T}$ be a commutative semigroup of continuous mappings of X into itself. Then the set $\tilde{P}_1$ of $\mathcal{T}$-invariant probability measures on X is nonempty, and for each $\mu \in \tilde{P}_1$ there exists a probability measure m on $\tilde{P}_1$ which is supported by the ergodic measures and which satisfies*

$$\mu(f) = \int \lambda(f) \, dm(\lambda) \quad \text{for each } f \in C(X).$$

Proof: The application of the Markov-Kakutani fixed-point theorem described above is sufficient to guarantee that $\tilde{P}_1$ be nonempty. Since X is a metric space, the set P_1 of all probability measures on X is weak* compact and metrizable (Section 2) and hence the same is true of its weak* closed convex subset $\tilde{P}_1$. If $\mu \in \tilde{P}_1$, then the Choquet theorem yields a representing measure m for μ supported by $\text{ext}\,\tilde{P}_1$, the set of ergodic measures, and the weak* continuity of the affine map $\lambda \to \lambda(f)$ ($f \in C(X)$) gives us the formula above.

4. THE MINKOWSKI-CARATHÉODORY REPRESENTATION THEOREM AND DOUBLY STOCHASTIC MATRICES

In this brief section we will sketch a proof of Carathéodory's version of Minkowski's theorem and describe one of its classical applications.

THEOREM (Carathéodory): *Suppose that K is a compact convex subset of $\mathbf{R}^n$. Then each point of K can be expressed as a convex combination of at most $n+1$ extreme points of K.*

Proof: We proceed by induction. The result is obvious if $n=1$, since in this case K is a closed line segment (possibly degenerate).

Suppose $n=2$, that K is not a segment, and that $x \in K$. If x is in the boundary of K, then there is a line L in $\mathbf{R}^2$ passing through x such that K is contained in one of the closed half-spaces determined by L. The intersection of L and K is necessarily a closed line segment (possibly degenerate) whose endpoints (=extreme points) are extreme points of K. Since the theorem is true for one-dimensional sets, we can express x as a convex combination of at most two extreme points of K. If, on the other hand, x is in the interior of K, let y be any extreme point of K. (Such points exist, by the argument just given.) Draw the line determined by y and x; its intersection with K is a closed segment $[y,z]$ which contains x in its relative interior and whose other endpoint z is naturally a boundary point of K. By what was proved above, z is a convex combination $z = \alpha u + (1-\alpha)v$ of two (or fewer, if z is extreme) extreme points of K. Since $x = \beta z + (1-\beta)y$, we can write

$$x = \beta \alpha u + \beta(1-\alpha)v + (1-\beta)y$$

and this is a convex combination of extreme points. The general induction step is carried out in exactly the same manner, replacing the line L by a hyperplane which supports K at the boundary point x.

Note that the number $n+1$ cannot be reduced; in the case $n=2$, for example, any interior point of a triangle requires exactly three extreme points in its representation.

DEFINITION: A real $n \times n$ matrix A is said to be *doubly stochastic* if each of its entries a_{ij} is nonnegative and if the sum of the entries in each row and the sum in each column is exactly 1.

Such matrices arise naturally in linear algebra as well as in probability and statistics, and in the study of inequalities.

An $n \times n$ matrix A can be considered, in an obvious way, as an element of $\mathbf{R}^{n^2}$; simply write down the rows of A one after another, say, to get the corresponding n^2-tuple. Let K be the subset of $\mathbf{R}^{n^2}$ corresponding to the set of all such matrices; it is clear that K is convex and (since the topology of $\mathbf{R}^{n^2}$ is given by pointwise convergence) closed. Moreover, the sum of the coordinates of any point in K is n. Since each coordinate is between 0 and 1, this implies that the Euclidean norm of such a point is at most $\sqrt{n}$. This shows that K is bounded, hence it is compact and Carathéodory's theorem applies. It remains to identify the extreme points of K.

PROPOSITION 4.1: *The extreme doubly stochastic matrices are the permutation matrices, i.e., those whose entries are either 0 or 1.*

The permutation matrices, then, are those which have exactly one entry 1 in each row and each column, with the remaining entries being 0.

One part of the proof is easy: Suppose that $A = (a_{ij})$ is a permutation matrix and that $A = \frac{1}{2}(B + C)$, where $B = (b_{ij})$ and $C = (c_{ij})$ are doubly stochastic. We then have

$$a_{ij} = \tfrac{1}{2}(b_{ij} + c_{ij})$$

for each i and j and hence $b_{ij} = a_{ij} = c_{ij}$ whenever $a_{ij} = 0$. If $a_{ij} = 1$, then $b_{ij} + c_{ij} = 2$ and therefore both of these are 1, which proves that $B = A = C$ and hence that A is extreme.

There are many proofs of the harder implication. An excellent discussion and survey may be found in [11].

The above results combine to yield the following: *Every doubly stochastic matrix is a convex combination of at most $n^2 + 1$ permutation matrices.* Actually, a much smaller number will suffice, since the set K of doubly stochastic matrices actually has dimension less than n^2. This is because the definition of K requires that the rows and columns of each matrix add up to 1, so that, for instance, the affine functional

$$(a_{ij}) \mapsto a_{11} + a_{12} + \cdots + a_{1n} - 1$$

vanishes on K. A simple argument (see [11]) shows that among the $2n$ such functionals any $2n-1$ are linearly independent, so that K has dimension at most $n^2-(2n-1)=(n-1)^2$. This implies that we can replace n^2+1 by $(n-1)^2+1$ in the above statement and it can be shown that the latter number is the best possible.

5. TWO TOOLS: THE SEPARATION THEOREM AND THE RESULTANT MAP

Probably the single most important tool in the study of convexity is the *separation theorem*, which asserts that if C is a closed convex subset of a Banach space E (or, more generally, of a locally convex Hausdorff linear space), and if $x \notin C$, then there exists $f \in E^*$ such that $\sup\{f(y): y \in C\} < f(x)$. [The word "separation" comes from the fact that for any number α between the two numbers above, the closed hyperplane $f^{-1}(\alpha)$ separates E into two half-spaces, one containing C, the other containing x.] This theorem is a geometric reformulation of a general version of the Hahn-Banach theorem and is not hard to prove, once the appropriate definitions have been introduced. For our purposes, however, a more special version suffices and its proof is almost immediate.

THEOREM 5.1 (Separation theorem in Hilbert space): *Let K be a nonempty compact convex subset of the Hilbert space H and suppose $x \notin K$. Then there exists $f \in H^*$ such that*

$$\sup\{f(y): y \in K\} < f(x).$$

Proof: Since K is compact, the continuous function $y \mapsto \|y-x\|$ attains its strictly positive minimum at some point $y \in K$. Let $z = x - y$ and for $u \in H$ let $f(u) = (u, z)$; this is clearly linear and continuous. The convexity of K implies that if $u \in K$ then $u_\lambda = \lambda u + (1-\lambda)y \in K$ whenever $0 < \lambda < 1$. Consequently,

$$0 < \|z\|^2 = \|x-y\|^2 \leq \|x-u_\lambda\|^2 = \|z - \lambda(u-y)\|^2.$$

By expanding the right side of this inequality, canceling $\|z\|^2$ and dividing by λ, we obtain

$$2(u-y,z) \leq \lambda \|u-y\|^2.$$

If we then take the limit of this as $\lambda \to 0$, we see that the left side is bounded above by 0; hence

$$f(u) = (u,z) \leq (y,z) = (x-z,z) = (x,z) - \|z\|^2 = f(x) - \|z\|^2.$$

Since this is true for all $u \in K$ we conclude that

$$\sup\{f(u): u \in K\} \leq f(x) - \|z\|^2 < f(x),$$

which was to be proved.

An examination of the proof above shows that we only used the compactness of K to ensure that it contained the point y of minimum distance from x. Such a point exists even when K is only assumed to be closed and convex. Although we do not use the more general result, it is interesting enough to warrant a sketch of its proof. By translation, it suffices to show that any such set K contains a point of smallest norm. Suppose, then, that $d = \inf\{\|x\|: x \in K\}$ and choose a sequence $\{x_n\} \subseteq K$ such that $\|x_n\| \to d$; we need only show that this sequence is Cauchy and hence convergent. By expanding the right side of $\|x \pm y\|^2 = (x \pm y, x \pm y)$ and adding the resulting identities we get the "parallelogram law":

$$\|x-y\|^2 + \|x+y\|^2 = 2(\|x\|^2 + \|y\|^2), \qquad x, y \in H.$$

Now, for any n and m we have $\frac{1}{2}(x_n + x_m) \in K$ so that $\|x_n + x_m\| \geq 2d$ and hence

$$\|x_n - x_m\|^2 \leq 2(\|x_n\|^2 + \|x_m\|^2) - 4d^2.$$

Since the limit of the right side as $n, m \to \infty$ is 0, we see that $\{x_n\}$ is a Cauchy sequence and the proof is complete.

The Choquet theorem shows that every point of a compact convex subset K of H is the resultant (or barycenter) of some probability measure on $\text{ext}\, K$. It does *not* imply, however, that every probability measure on ext K (or on K) has a resultant. The following useful proposition shows that this is indeed the case; its proof uses the separation theorem in $\mathbf{R}^n$.

PROPOSITION 5.2: *Let $K \subseteq H$ and suppose that μ is a probability measure on K. Then there exists a unique point $x_\mu \in K$ such that $f(x_\mu) = \int_K f d\mu$ for each $f \in A(K)$.*

Proof: For each $f \in A(K)$ let $H_f = \{x \in K: f(x) = \mu(f)\}$. Each such set H_f is closed in K and a moment's thought shows that it suffices to prove that the intersection of all of these sets consists of a single point x_μ. By compactness, we can conclude that the intersection is nonempty once we have shown that this family of sets has the finite intersection property. Suppose, then, that $f_1, f_2, \ldots, f_n$ are in $A(K)$ and define $T: K \to \mathbf{R}^n$ by setting $T(x) = (f_1(x), f_2(x), \ldots, f_n(x))$; it suffices to show that $p \in T(K)$, where $p = (\mu(f_1), \mu(f_2), \ldots, \mu(f_n))$. Now, the map T is affine and continuous, so $T(K)$ is compact and convex. If $p \notin T(K)$, then there must exist an element $a = (a_1, a_2, \ldots, a_n)$ in $\mathbf{R}^n$ such that $(a, p) > \sup\{(a, T(x)): x \in K\}$. Define $f \in A(K)$ by $f = \sum_1^n a_k f_k$; then for $x \in K$ we have

$$(a, T(x)) = \sum_1^n a_k f_k(x) = f(x)$$

and

$$(a, p) = \sum_1^n a_k \mu(f_k) = \mu(f),$$

so the above inequality becomes

$$\int_K f d\mu > \sup\{f(x): x \in K\}.$$

This contradiction shows that $p \in T(K)$. Thus, the family of closed sets $\{H_f: f \in A(K)\}$ has the finite intersection property and therefore there is at least one point x in their intersection. If there were another such point y, we would have $f(x) = \mu(f) = f(y)$ for every

$f \in A(K)$, contradicting the fact that $A(K)$ separates the points of K. This completes the proof. (Note that we didn't actually use the fact that K is a subset of a Hilbert space, only that it is compact and that $A(K)$ separates points of K.)

As we noted in Section 1, the point x_μ obtained above is called the *resultant* of μ. What we have done is to define a map $\mu \mapsto x_\mu$ from $P_1(K)$ onto K, called the *resultant map*. It is simple to check that this map is affine. The Choquet representation theorem can be reformulated to say that, even if we restrict the resultant map to the set $P_1(\text{ext } K)$ of all probability measures on ext K, then it is still a surjective map.

6. UNIQUENESS OF THE CHOQUET REPRESENTATION AND SIMPLEXES

To guide us in formulating a criterion for uniqueness of the Choquet representation, we return to the finite dimensional situation. It is clear, for instance, that if a compact convex subset K of the plane has more than three extreme points, then we can find a point in K which can be expressed in two distinct ways as a convex combination of members of ext K. (In fact, any four distinct extreme points will form the vertices of a quadrilateral in K and the intersection of its diagonals is such a point.) Thus, if every point of K can be expressed in exactly one way as a convex combination of extreme points, then K must be a triangle. The converse assertion is equally obvious: Any point of a triangle has a unique representation as a convex combination of vertices (the coefficients being the well-known "barycentric coordinates"). The analogous result is valid in three-dimensional space, if we use tetrahedra in place of triangles. More generally, in an n-dimensional space E we simply replace triangles by "simplexes." The definition of the latter is simple. First, recall that a finite set $\{x_0, x_1, \ldots, x_n\}$ is *affinely independent* provided the equalities

$$\sum_0^n \alpha_k x_k = 0 \quad \text{and} \quad \sum_0^n \alpha_k = 0$$

can hold only if each of the real numbers $\alpha_0, \alpha_1, \ldots, \alpha_n$ is zero. (This is equivalent to saying that none of the points is an affine combination of the others, an *affine combination* being a linear combination in which the sum of the coefficients is 1.) A subset K of E is a *simplex* if it is the convex hull of $n+1$ affinely independent points.

The "standard" simplexes are perhaps the easiest to visualize: In $\mathbf{R}^{n+1}$, let $e_0=(1,0,0,\ldots,0)$, $e_1=(0,1,0,\ldots,0),\ldots,e_n=(0,\ldots,0,1)$ be the usual basis vectors and let S_n be the (n-dimensional) convex hull of these vectors. Thus, $x=(x_0,\ldots,x_n)\in S_n$ if and only if $x_k\geqslant 0$ for each k and $\sum_{k=0}^{n}x_k=1$. We could just as well have defined a finite dimensional convex set K to be a *simplex* provided there exists a one-one affine map from K onto some S_n. (This would, of course, be the map which assigns to each point of K its $(n+1)$-tuple of barycentric coordinates.) Now, the "positive orthant" $\mathbf{R}_+^{n+1}$ in $\mathbf{R}^{n+1}$ is the *cone generated by* S_n, that is, a point of $\mathbf{R}^{n+1}$ is in $\mathbf{R}_+^{n+1}$ (that is, has nonnegative coordinates) if and only if it is of the form λx for some $\lambda\geqslant 0$ and some $x\in S_n$. [If $y\neq 0$ is in $\mathbf{R}_+^{n+1}$, we take λ to be the sum of its coordinates and let $x=\lambda^{-1}y$ to get a representation of the above form.] The set $\mathbf{R}_+^{n+1}$ has an obvious property: If $x,y\in\mathbf{R}_+^{n+1}$, then so is the point z whose kth coordinate is given by $z_k=\max\{x_k,y_k\}$. This trivial observation is the key to extending the notion of "simplex" to infinite dimensional spaces. To do this properly, we first consider the notions of cones and their induced partial orderings in an arbitrary real vector space E.

DEFINITION: A subset P of a real vector space E is called a *cone* (with vertex at the origin) if it is convex and if $\lambda x\in P$ whenever $\lambda\geqslant 0$ and $x\in P$. We say that P is *proper* provided $P\cap(-P)=\{0\}$; that is, $x=0$ is the only point for which both x and $-x$ are in P.

The set $\mathbf{R}_+^{n+1}$ is a proper cone. More generally, if X is any compact Hausdorff space, then the nonnegative continuous functions on X form a proper cone in $C(X)$. The set $P(X)$ of all nonnegative finite regular Borel measures on X is a proper cone, which can also be described as the set of all nonnegative linear functionals on $C(X)$. An interesting example of a cone which is *not* proper is the cone S of all continuous convex functions on the compact convex subset K of Hilbert space. In this instance, we have $S \cap (-S) = A(K)$.

A cone $P \subseteq E$ always induces a partial ordering on E by defining:

$$x \geqslant y \quad \text{means} \quad x - y \in P.$$

It is easily verified that such an ordering is transitive. It is also proper ($x \geqslant y$ and $y \geqslant x$ implies $x = y$) provided P is proper. Moreover, the ordering is translation invariant, in the sense that

$$x \geqslant y \quad \text{if and only if} \quad x + z \geqslant y + z \quad \text{for all } z \in E,$$

and it also satisfies

$$\lambda x \geqslant \lambda y \quad \text{whenever} \quad x \geqslant y \quad \text{and} \quad \lambda \geqslant 0.$$

Conversely, given a partial ordering on E with all the above properties, the set $P = \{x \in E : x \geqslant 0\}$ is a proper convex cone which induces the given ordering. (When there are two such orderings on a space, one must be careful to distinguish them. For instance, in the ordering induced on $A(K)$ by the cone C of convex functions, to say that $f \geqslant g$ means that $f - g$ is convex; it need not be a nonnegative function.)

If P is a convex cone and $x, y \in P$, then there is always an element $z \in P$ such that $z \geqslant x$ and $z \geqslant y$ (for example, we can take $z = x + y$). Such an element is called an *upper bound* for x and y. We call z the *least upper bound of x and y* provided z is an upper bound for x and y and $z \leqslant w$ whenever w is an upper bound for x and y. The least upper bound for x and y is necessarily unique (when P is proper) and is denoted by $x \vee y$. For example, if $P = \mathbf{R}_+^{n+1}$, then $x \vee y$ is the element whose kth coordinate is the maximum of x_k and y_k. In an arbitrary proper cone P, least upper bounds need not exist. We will see a concrete example shortly, but a geometrical interpretation of $x \vee y$ shows this fairly clearly: The set of elements of E which are greater or equal to x is the translate of P to the cone $P + x$ with vertex x. Similarly, $P + y$ is the cone of elements which dominate y and hence $B = (P + x) \cap (P + y)$ is the set of all upper bounds for x and y. This set B will have a least element z if and only if $w \geqslant z$ whenever $w \in B$, that is, if and only if $w - z \in P$ whenever $w \in B$. This last assertion is the same as saying that $w \in P + z$ if and only if $w \in B$, which is equivalent to $P + z = (P + x) \cap (P + y)$. Now, it is easy to find cones in $\mathbf{R}^3$ for which this need not be the case; for instance, if P is a cone with a square cross-section. On the other hand, if P has a triangular cross-section, then the intersection of any two translates is again a translate of P. Both situations are shown in Figure 2 where, for the sake of clarity, the cones $P + x$ and $P + y$ have been truncated by a single plane.

If $x \vee y \in P$ whenever $x, y \in P$, then we say that P is *lattice ordered*. As the last example suggests, there is a close connection between certain lattice ordered cones and simplexes. To see how this comes about, we need some additional definitions.

If K is a convex subset of a linear space, with $0 \notin K$, then the set

$$\mathbf{R}^+ K = \{\lambda x : \lambda \geq 0, x \in K\}$$

is a proper convex cone, called the *cone generated* by K. In order that the properties of this cone accurately reflect those of K, we need to assume that K is contained in a hyperplane which misses the origin; that is, that there exists a nontrivial linear functional f on the space such that $K \subseteq f^{-1}(\alpha)$ for some real number $\alpha \neq 0$. (Otherwise, radically different convex sets, even in the plane, can generate the same cone.) We will say that K is a *base* for the cone P if there exists a hyperplane J which misses the origin such that $K = P \cap J$. It is elementary to verify that if K is a base for P, then P is generated by K. A canonical way to express an arbitrary convex set as the base of a cone is the following: If $K \subseteq E$, let $E_1 = E \times \mathbf{R}$; this product space is a linear space, with addition and scalar multiplication defined in the obvious way. Let $K_1 = K \times \{1\} = \{(x, 1) : x \in K\} \subseteq E_1$. Then K_1 is contained in the hyperplane $H_1 = \{(x, r) \in E_1 : r = 1\}$ and the map $T : K \to K_1$ defined by $T(x) = (x, 1)$ is an affine bijection of K onto K_1. Moreover, if E is a normed linear space, say, and E_1 is given the product topology, then T is a homeomorphism.

It is easy to verify that an affine one-one map T from the base K_1 of one cone P_1 onto the base K_2 of another cone P_2 can be extended (by $T'(\lambda x) = \lambda T(x)$, $x \in K_1$, $\lambda \geq 0$) to an affine one-one

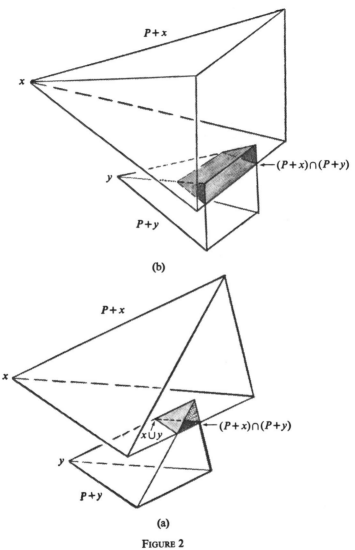

FIGURE 2

map T' of P_1 onto P_2 which preserves order; hence P_1 will be lattice ordered if and only if the same is true of P_2. This observation allows us to make the following definition.

DEFINITION: A convex set K in a linear space E is said to be a *simplex* provided it is the one-one affine image of the base of a lattice ordered cone.

It is clear from the definition of a finite dimensional compact simplex and the observation that $\mathbf{R}_+^{n+1}$ is a lattice-ordered cone with base S_n that any finite dimensional compact simplex is a simplex in the sense of the Definition. It is also true, but harder to prove (for example, see [12]), that a finite dimensional compact convex set which satisfies the definition above is affinely homeomorphic to S_n. Thus, the definition above is a reasonable one to use in infinite dimensional spaces. We will usually apply it to *compact* convex sets. An example is the set $K = P_1(X)$ of all regular Borel probability measures on the compact Hausdorff space X; we know that K is weak* compact convex, and it is clearly a base for the cone $P(X)$ of all nonnegative measures. We must show, of course, that $P(X)$ is lattice ordered. More generally, we have the following result.

PROPOSITION 6.1: *Let X be a set, Σ a σ-algebra of subsets of X and let P be the cone of all finite positive measures on the measurable space (X, Σ). Then P is a lattice ordered cone in the space M of all finite signed measures on (X, Σ), with base P_1 consisting of the probability measures on (X, Σ). Thus, the probability measures on any measure space form a simplex.*

Proof: It is clear that P_1 is contained in the hyperplane $\{\mu \in M: \mu(X) = 1\}$ and that P_1 is a base for P. Note that the ordering defined by P is the natural one: By definition, to say that $\lambda \geqslant \mu$ in this ordering is the same as saying that $\lambda - \mu \in P$, that is, that $\lambda(B) \geqslant \mu(B)$ for each measurable set B. Now, given $\mu_1, \mu_2 \in P$, we can find $\mu_1 \vee \mu_2$ as follows: Let $\mu = \mu_1 + \mu_2$; then both μ_1, μ_2 are absolutely continuous with respect to μ and hence, by the Radon-Nikodým theorem, there exist measurable functions f_1, f_2 on X

such that $\mu_k = f_k \mu$, $k = 1, 2$. Let f be defined by $f(x) = \max\{f_1(x), f_2(x)\}$ for $x \in X$; then f is measurable and μ-integrable and it is straightforward to verify that $f\mu = \mu_1 \vee \mu_2$.

Choquet's uniqueness theorem is the following:

THEOREM 6.2 (Choquet): *Let K be a compact convex subset of a separable Hilbert space H. Then to each point $x \in K$ there corresponds a unique representing probability measure μ on* ext$K$ *if and only if K is a simplex.*

The proof that uniqueness holds for simplexes requires somewhat more machinery than we wish to include here. A succinct proof (which includes the nonmetrizable case) may be found in [1, p. 87]. We can, however, prove the converse implication with no additional effort. Indeed, if we assume that every point of K is the resultant of exactly one probability measure on ext$K$, then the restriction of the resultant map to $P_1(\text{ext}\,K)$ is one-one and surjective. Since, by Proposition 6.1, the set $P_1(\text{ext}\,K)$ is a simplex, the uniqueness assumption thus shows that K is the one-one affine image of a simplex, hence is itself a simplex.

There is another description of simplexes which is not as useful as the "lattice ordered cone" definition, but which has considerable geometrical appeal. Suppose that K is a compact convex subset of $\mathbf{R}^2$, say. If $\alpha > 0$, then αK is also a compact convex set which is not only similar to K but is oriented the same way, since multiplication by a positive scalar does not cause any rotation. If we then apply a translation by a fixed vector x, we get $\alpha K + x$, a *homothetic image* of K. Given two such homothetic images, their intersection $(\alpha K + x) \cap (\beta K + y)$ will either be empty or another compact convex set. A few minutes with pencil and paper will persuade one of the validity of the following two assertions: If K is a *triangle* and if the intersection above is more than a single point, then it will be of the form $\gamma K + z$ for some $\gamma > 0$, $z \in \mathbf{R}^2$. (This is illustrated in Figure 2.) Moreover, if K is *not* a triangle, there will always exist a homothetic image of K whose intersection with K is nontrivial but is not itself a homothetic image of K. More generally, the same two assertions hold in $\mathbf{R}^n$ if "triangle" is replaced by "simplex," so we have a different kind of characterization of finite dimensional simplexes. The following theorem, which we state without proof, shows again that the definition of "simplex" in the infinite dimensional situation is amply justified.

THEOREM (Choquet): *Let K be a compact convex subset of H (or any Hausdorff linear space). Then K is a simplex if and only if the intersection of any two homothetic images of K is either empty, a single point, or a homothetic image of K.*

A proof may be found in [10].

As an application of the uniqueness theorem we return to the example of invariant and ergodic measures and show that the representation obtained in that case is always unique. To do this, we will show that the cone $P = \mathbf{R}^+ \tilde{P}_1$ generated by the set $\tilde{P}_1$ of invariant probability measures is lattice ordered. Obviously, P is the set of all positive invariant measures. We again borrow a lemma from [12, Section 10], and we use the notation from Section 3.

LEMMA 6.3: *Suppose that λ is a positive invariant measure and that μ is a positive measure on Ω which is absolutely continuous with respect to λ, and with Radon-Nikodým derivative f. Then μ is an invariant measure if and only if $f = f \circ T$ a.e. λ, for all T in $\mathfrak{T}$.*

The sufficiency half of this lemma is easy: If $f = f \circ T$ a.e. λ, then for any measurable set A we have

$$\mu(T^{-1}A) = \int_{T^{-1}A} f \, d\lambda = \int_{T^{-1}A} f \circ T \, d\lambda$$
$$= \int_A f \, d(\lambda \circ T^{-1}) = \int_A f \, d\lambda = \mu(A).$$

THEOREM 6.4: *The cone P of all finite nonnegative $\mathfrak{T}$-invariant probability measure is a lattice in its own ordering.*

Proof: As is common in proofs of this kind, we must be aware of the difference between two natural orderings; the one induced by P and the one induced by the cone of all positive measures. To say that μ is greater than λ in the latter sense is the same as saying that $\mu - \lambda$ is a positive measure, while the former requires $\mu - \lambda$ to be a positive *invariant* measure. In this case we are fortunate, since if μ and λ are invariant, then so is $\mu - \lambda$ and hence positivity of the latter measure suffices in either case. Suppose, now, that μ, ν are two measures in P; we want to show that they have a least upper bound in P. Let $\lambda = \mu + \nu$; then μ and ν are both absolutely continuous with respect to λ and the latter is invariant, so by Lemma 6.3 their Radon-Nikodým derivatives f, g with respect to λ satisfy

$$f = f \circ T \quad \text{a.e. } \lambda, \quad \text{and} \quad g = g \circ T \quad \text{a.e. } \lambda,$$

for all T in $\mathcal{T}$. It follows that the function $h = \max(f, g)$ also satisfies $h = h \circ T$ a.e. λ, and (as shown in the proof of Proposition 6.1) the measure $h\lambda$ is the least upper bound of μ and ν for the usual ordering on measures. From the remarks above it follows that $h\lambda$ is a least upper bound in P.

The theorem above shows, then, that whenever we can get a representation theorem for an arbitrary invariant probability measure as an integral average of ergodic measures, then the representation is unique. In particular, we can deduce the following theorem from the material discussed at the end of Section 3.

THEOREM 6.5: *Let X be a compact metric space and let $\mathcal{T}$ be a commutative semigroup (under composition) of continuous maps of X into itself. Then the set $\tilde{P}_1$ of $\mathcal{T}$-invariant regular Borel probability measures on X is nonempty, and to each such measure μ there corresponds a unique probability measure m on the ergodic measures in $\tilde{P}_1$ such that*

$$\mu(f) = \int_{\tilde{P}_1} \lambda(f) \, dm(\lambda) \quad \text{for } f \in C(X).$$

7. THE KREIN-MILMAN THEOREM AND RELATED TOPICS

As was remarked in the introduction, the Choquet theorem is not the only possible extension of Minkowski's theorem to infinite dimensional spaces. Whereas the Choquet theorem preserves the idea of expressing an arbitrary point of a compact convex set K as a barycenter of (possibly uncountably many) extreme points of K, the Krein-Milman theorem involves only finite convex combinations of extreme points; it asserts that such convex combinations are *dense* in K. The proof of the Krein-Milman theorem, even for the most general case of a compact convex subset of a locally convex Hausdorff linear space, is not difficult. It is amusing and instructive, however, to see that it follows easily from Choquet's theorem. Since we have only proved the latter when K is a subset of a separable Hilbert space H, we restrict ourselves to this case. [As remarked earlier, this actually covers the case of a metrizable compact convex subset of a locally convex space.]

THEOREM 7.1 (Krein-Milman): *Let K be a compact convex subset of H and let K_0 be the closure of the set of all finite convex combinations of points from $\text{ext}\, K$. Then $K_0 = K$.*

Proof: It is easily seen that the set of all convex combinations of points in $\text{ext}\, K$ is convex and that its closure K_0 is also convex. Since K_0 is a subset of K, it is compact. We will obtain a contradiction to the Choquet theorem by assuming that K_0 is a proper subset of K. Without loss of generality, then, we assume that the origin 0 belongs to $K \setminus K_0$. Since K_0 is compact, the separation theorem of Section 5 guarantees the existence of a continuous affine function h on K which satisfies $\sup\{h(x): x \in \text{ext}\, K\} < h(0)$. By Choquet's theorem, we can find a probability measure μ on $\text{ext}\, K$ with barycenter 0, so that μ satisfies

$$\int_K f(x)\, d\mu(x) = f(0)$$

for any continuous affine function f on K. But since $\mu(\text{ext}\, K) = 1$, this implies that

$$h(0) = \int_K h(x)\,d\mu(x)$$
$$\leq \mu(\operatorname{ext} K)\cdot \sup\{h(x): x \in \operatorname{ext} K\}$$
$$< h(0),$$

which is a contradiction.

DEFINITION: If A is a subset of a normed linear space (or, more generally, of a topological vector space), we define the *convex hull* of A to be the intersection of all convex sets containing A, and we denote it by co A. Since, by definition, co A is the smallest convex set containing A, it coincides with the set of all convex combinations of points from A. The closure $\overline{\operatorname{co}}\, A$ of co A is called the *closed convex hull* of A; it coincides with the intersection of all closed convex sets containing A.

The Krein-Milman theorem can be reformulated to state that $K = \overline{\operatorname{co}}(\operatorname{ext} K)$ for any compact convex subset K of H. More generally, for any $A \subseteq K$, if ext $K \subseteq A$, then $\overline{\operatorname{co}}\, A = K$. A converse to this latter statement would be the assertion that if $\overline{\operatorname{co}}\, A = K$, then ext $K \subseteq A$. This is easily seen to be false; indeed, let K be a closed circular disk in the plane and let A be the boundary circle with one point deleted. However, if A is *closed*, we do get a partial converse, an interesting and frequently useful result due to Milman. It can be proved in several ways; the proof we present will utilize some results concerning barycenters of measures which are interesting and useful in themselves. The first of these characterizes certain closed convex hulls in terms of representing measures, while the second characterizes extreme points in similar terms.

PROPOSITION 7.2: *Let K be a compact convex subset of H and let A be a subset of K. Then $\overline{\operatorname{co}}\, A = K$ if and only if for each $x \in K$ there is a probability measure μ on the closure of A with barycenter x.*

Proof: Suppose, first, that $\overline{\text{co }} A = K$; then by definition, there is a sequence $\{y_n\}$ in co A (that is, a sequence of convex combinations of points of A) which converges to x. To each such convex combination y_n there corresponds (as in Section 1) a discrete measure μ_n on A with barycenter y_n. Let X denote the closure of A. By Proposition 2.1, the set of all probability measures on X is compact and metrizable in the weak* topology induced by $C(X)$, so there exists a subsequence (call it again $\{\mu_n\}$) which converges weak* to a probability measure μ on X. If $f \in A(K)$, then the restriction of f to X is continuous, so that $\mu_n(f) \to \mu(f)$. Since f is affine, we have $\mu_n(f) = f(y_n)$ and since $y_n \to x$, we also have $f(y_n) \to f(x)$, which shows that x is the barycenter of μ.

To prove the converse, we copy the proof above of the Krein-Milman theorem, replacing K_0 by $\overline{\text{co }} A$ and using the assumed existence of a representing measure on X in place of the Choquet theorem.

PROPOSITION 7.3 (Bauer): *Let K be a compact convex subset of H. Then $x \in \text{ext } K$ if and only if the only probability measure μ on K with barycenter x is the point mass ϵ_x.*

Proof: If $x \in K$ is not extreme, then it is a nontrivial convex combination of distinct points of K and the corresponding discrete measure μ has barycenter x. Since $\mu \neq \epsilon_x$, this completes half of the proof. Suppose, now, that $x \in \text{ext } K$ and that the probability measure μ on K has barycenter x. We want to show that $\mu = \epsilon_x$, that is, that the closed support supp μ of μ is equal to $\{x\}$. (Recall that $y \in \text{supp } \mu$ if and only if $\mu(U) > 0$ for every open subset U which contains y.) Now, if there exists $y \in \text{supp } \mu$, $y \neq x$, then we can find $f \in A(K)$ and $r \in \mathbf{R}$ such that $f(y) < r < f(x)$. The compact set $B_0 = \{z \in K : f(z) \leqslant r\}$ contains y in its relative interior and therefore has positive μ measure, equal to $m > 0$, say. If $m = 1 = \mu(B_0)$, then

$$f(x) = \int f d\mu = \int_{B_0} f d\mu \leqslant r < f(x),$$

a contradiction. Thus, the two measures μ_1, μ_2 defined for any

Borel set B by

$$\mu_1(B) = m^{-1}\mu(B \cap B_0) \quad \text{and} \quad \mu_2(B) = (1-m)^{-1}\mu(B \cap (K \setminus B_0)),$$

are probability measures on K, with resultants which we denote by x_1, x_2, respectively. It is easy to verify that $\mu = m\mu_1 + (1-m)\mu_2$, which (since the resultant map is affine) implies that $x = mx_1 + (1-m)x_2$. We know that x is an extreme point and that $0 < m < 1$, so we must have $x = x_1 = x_2$. On the other hand, B_0 is compact and convex and μ_1 is actually a probability measure on B_0, so by Proposition 5.2 its resultant x_1 is in B_0. But this implies that $f(x) = f(x_1) \leq r$, a contradiction which completes the proof.

The proof of Milman's theorem is now quite easy.

THEOREM 7.4 (Milman): *If K is a compact convex subset of H and A is a closed subset of K such that $\overline{\text{co}}\, A = K$, then $\text{ext}\, K \subseteq A$.*

Proof: If $x \in \text{ext}\, K \subseteq K$, then $x \in \overline{\text{co}}\, A$, so by Proposition 7.2 there is a probability measure μ on A with barycenter x. By Proposition 7.3 we must have $\mu = \epsilon_x$ and therefore $x \in A$.

Obviously, an equivalent reformulation of Milman's theorem is the assertion that if $B \subseteq K$ and $\overline{\text{co}}\, B = K$, then $\text{ext}\, K$ is contained in the closure of B.

We can also reformulate the Krein-Milman theorem as an integral representation theorem; this is another easy consequence of Proposition 7.2.

THEOREM 7.5 (Krein-Milman): *Suppose that K is a compact convex subset of H and let A denote the closure of the extreme points of K. Then for each x in K there exists a probability measure μ on A with barycenter x.*

Proof: (What we will prove is that the Krein-Milman theorem implies the above assertion, and conversely.) By the Krein-Milman theorem, $K = \overline{\text{co}}(\text{ext}\, K)$ and by Proposition 7.2 we can get a representing measure on A for each $x \in K$. Conversely, if such measures exist, then by Proposition 7.2 again, we can conclude that $K = \overline{\text{co}}(\text{ext}\, K)$, which is the Krein-Milman theorem.

The result above clearly shows how the Choquet theorem sharpens the Krein-Milman theorem, since the former always gives us a representing measure on the extreme points, rather than on their closure. That this is a significant improvement is best illustrated by an example due to Poulsen [14]: There exists a compact convex simplex $K \subseteq H$ such that ext K is *dense* in K. (For any point x in this set, then, the point mass ϵ_x gives a representing measure which sits on the closure of ext K, hence the "trivial representation" is as good as that provided by the Krein-Milman theorem.)

8. POSSIBLE GENERALIZATIONS AND SUGGESTIONS FOR FURTHER READING

The Choquet existence and uniqueness theorems and the Krein-Milman theorem which we have presented for a compact convex subset K of Hilbert space are (as indicated several times) actually valid for any metrizable compact convex subset of a locally convex Hausdorff topological vector space. Aside from this straightforward extension (the same proofs go through), how much further can these results be generalized? First, it has recently been shown by J. W. Roberts [15] that there exists a Hausdorff topological vector space which contains a nonempty compact convex set with no extreme points. Since the existence theorems imply, in particular, that K has extreme points, this shows that the hypothesis of local convexity cannot be discarded without adding some additional assumption. What can be said about the other main hypotheses, metrizability and compactness? The Krein-Milman theorem is valid without metrizability, but the Choquet theorem is not. The problem arises from the requirement that a representing measure assign mass 1 to the set of extreme points: In the nonmetrizable case, it is possible for the extreme points to form a non-Borel set. One *can* hope for measures which "almost" sit on the extreme points, however, and in this weaker sense, the theorem is true for arbitrary compact convex sets in a locally convex space. For instance, in proving Choquet's theorem, we obtained a measure μ which satisfies $\mu(f) = \mu(\bar{f})$ for a strictly convex continuous function f, and this was enough to ensure that μ sat on ext K. In the general case it is possible to get representing measures which satisfy the above equality for *every* convex continuous function f, and even though there exists no *strictly* convex continuous f, this still turns out to be quite useful. In particular, the uniqueness theorem is valid in the nonmetrizable case if one uses measures with the above property. These matters are treated in detail in several books [1], [3], [12], [13].

The closed unit ball $K = \{x: \|x\| \leq 1\}$ of the space c_0 of all real sequences $x = (x_n)$ which converge to zero (with $\|x\| = \sup|x_n|$) is a classical example of a separable, bounded closed convex subset of a Banach space with ext K empty, so it is hopeless to try to eliminate compactness from either the Choquet or Krein-Milman theorems. There is, however, a special class of Banach spaces in which the Choquet existence and uniqueness theorems are valid for any separable bounded closed convex subset. This class arises in connection with vector-valued versions of the Radon-Nikodým theorem and it includes all reflexive Banach spaces and separable dual Banach spaces (such as l_1, the dual of c_0). These Choquet-type theorems are rather recent and are due to Edgar [8], [9], Bourgin and Edgar [2] and (independently) Saint Raymond [16]. The Krein-Milman theorem can also be considerably generalized in these "Radon-Nikodým" spaces; not only can "compact" be replaced by "closed and bounded," but the extreme points can be replaced by a smaller set (the so-called strongly exposed points). A detailed exposition may be found in [5, Chapter 6].

We conclude by describing an example of an integral representation theorem in the classical sense which can be formulated (and proved) as an integral representation theorem in our sense. Details may be found in [12, Section 2] or [3, Section 32].

DEFINITION: A real valued function f on $(0, \infty)$ is said to be *completely monotone* if it has derivatives $f^{(0)} = f, f^{(1)}, f^{(2)}, \ldots$ of all orders and if $(-1)^n f^{(n)} \geq 0$ for $n = 0, 1, 2, \ldots$.

A completely monotone function f is not only nonnegative and nonincreasing, but the same is true of each of the functions

$(-1)^n f^{(n)}$. Examples of such functions are $x^{-\alpha}$ and $e^{-\alpha x}$ ($\alpha \geqslant 0$). The set K of all completely monotone functions f for which $f(0^+) \leqslant 1$ is easily seen to be convex. Moreover, it is compact and metrizable in the topology of uniform convergence (of functions and all their derivatives) on compact subsets of $(0, \infty)$. The extreme points of K consist of the zero function and the functions $e^{-\alpha x}$, $\alpha \geqslant 0$, and the set ext K is closed in K. If f is a nonzero completely monotone function and if $f(0^+) < \infty$, then (dividing by $f(0^+)$) we can normalize f so that $f(0^+) = 1$. Since $f \in K$, the integral form of the Krein-Milman theorem (Theorem 7.5) shows that there exists a probability measure μ on ext K which represents f. It can be shown that μ is actually supported by the nonzero extreme points of K and is unique. By using the fact that the correspondence between $\alpha \in [0, \infty)$ and $e^{-\alpha x} \in \text{ext} K$ is a homeomorphism, one can carry μ to a measure λ on $[0, \infty)$. Finally, by using the fact that for each $x \in (0, \infty)$ the function $g \mapsto g(x)$, $g \in K$, is affine and continuous, we get the following classical integral representation theorem.

THEOREM 8.1 (S. Bernšteĭn): *If f is a completely monotone function on $(0, \infty)$ with $f(0^+) < +\infty$, then there exists a unique finite Borel measure λ on $[0, \infty)$ such that for each $x > 0$,*

$$f(x) = \int_0^\infty e^{-\alpha x} d\lambda(\alpha).$$

By differentiating under the integral sign it is easily seen that any function which can be represented by such an integral formula is completely monotone, so Bernšteĭn's formula actually characterizes the completely monotone functions.

REFERENCES

1. Erik M. Alfsen, *Compact Convex Sets and Boundary Integrals*, Ergebnisse der Math. (Band 57), Springer-Verlag, Berlin-Heidelberg-New York, 1971.
2. R. D. Bourgin and G. A. Edgar, "Noncompact simplexes in Banach spaces with the Radon-Nikodým property," *J. Functional Analysis*, 23 (1976), 162–176.
3. G. Choquet, *Lectures on Analysis*, Benjamin, New York, 1969.

4. Mahlon M. Day, *Normed Linear Spaces*, 3rd ed., Ergebnisse der Math. (Band 21), Springer-Verlag, Berlin-Heidelberg-New York, 1973.
5. Joseph Diestel, *Geometry of Banach Spaces—Selected Topics*, Lecture Notes in Math., No. 485, Springer-Verlag, Berlin-Heidelberg-New York, 1975.
6. J. Dieudonné, *Foundations of Modern Analysis*, Academic Press, New York-London, 1960.
7. N. Dunford and J. Schwartz, *Linear Operators, Part I: General Theory*, Interscience Publishers, New York-London, 1958.
8. G. A. Edgar, "A noncompact Choquet theorem," *Proc. Amer. Math. Soc.*, 48 (1975), 354–358.
9. _____, "Extremal integral representations," *J. Functional Analysis*, 23 (1976), 145–161.
10. D. G. Kendall, "Simplexes and vector lattices," *J. London Math. Soc.*, 37 (1962), 365–371.
11. L. Mirsky, "Results and problems in the theory of doubly-stochastic matrices," *Z. Wahrscheinlichkeitstheorie und Verw. Gebiete*, 1 (1963), 319–334.
12. Robert R. Phelps, *Lectures on Choquet's Theorem*, D. Van Nostrand, Princeton, N.J., 1966.
13. _____, *Lectures on Choquet's Theorem*, rev. ed., Ergebnisse der Math., Springer-Verlag, Berlin-Heidelberg-New York (to appear).
14. Ebbe Thue Poulsen, "A simplex with dense extreme points," *Ann. Inst. Fourier*, 11 (1961), 83–87.
15. James W. Roberts, "A compact convex set without extreme points," *Studia Math.*, 60 (1977), 255–266.
16. Jean Saint Raymond, "Représentation intégrale dans certains convexes," *Sém. Choquet, Initiation à l'Analyse*, 14e année, 1974/75, 11 pp.

ASPECTS OF BANACH LATTICES

H. H. Schaefer

INTRODUCTION

"Aspects of Banach Lattices" is a study of normed Riesz spaces and Banach lattices with a threefold purpose. Its first objective is to give the interested reader an independently readable introduction to the modern theory of Banach lattices. Second, beyond the scope of a totally elementary introduction it is designed to lead the reader to several recent non-trivial results of the theory. Finally, it presents a major application to cyclic Banach spaces which serves well to illustrate the bearing of a theory on an area of mathematics for whose benefit it was not developed.

Since no special introductions are given in the four Parts, we survey their contents briefly. Part I contains an introduction to the basic theory of normed Riesz spaces, but the material has been rigorously limited to the results indispensable for the subsequent Parts. Theorem I.4.4 on Banach lattices with order continuous norm is the only result of a non-introductory nature; partially based on a result of Part II (Dixmier's theorem), it gives the reader a taste of abstract Banach lattice theory.

Part II, discussing relations with function spaces, starts with the classical Kakutani-Krein [Kreĭn] representation of AM-spaces with unit, which constitutes a constantly used tool of the general theory. It then proceeds to the characterization of the L^p-spaces ($1 \leqslant p < \infty$) of an arbitrary measure space by the p-additivity of their norm, and on to the Dixmier-Grothendieck characterization of dual Banach lattices and Banach spaces $C(K)$. Finally, a representation of certain important types of Banach lattices as spaces of (continuous, extended real-valued) functions on a compact space is derived. While unifying and extending Kakutani's classical theorems, in many respects this representation is reminiscent of the Gelfand [Gel'fand] representation of commutative Banach algebras.

Part III presents an introduction to tensor products of Banach lattices and their applications. Since topological and even normed tensor products—though generally interesting, and indispensable for many parts of operator theory—are much neglected in functional analytic texts, an introductory section on the π-tensor product of normed spaces and its relevance for the extension of linear operators has been included. The subsequent section constructs a (Riesz space) tensor product of Archimedean Riesz spaces; the existence of such a tensor product is a recent, non-elementary result. A rather concise proof is given which is well suited to illustrate some of the methods of Riesz space theory. The usefulness of this tensor product for the theory of normed Riesz spaces is exemplified by the construction of the p-tensor product of Banach lattices as a predual for certain Banach lattices of operators, a result even more recent than the preceding one. Finally, all of this is applied to obtain extension theorems for positive operators, partly new (Theorem III.4.3) and partly discovered by different methods only a few years ago.

Part IV applies the results of Parts I and II to cyclic Banach spaces. A rather sweeping approach leads to many now classical results (and some apparently new ones), otherwise not easily accessible (cf. [5]), up to Bade's characterization of the strongly closed operator algebra generated by a σ-complete Boolean algebra of projections on a cyclic Banach space. Other areas of application of Banach lattice theory, such as ergodic theory, could have been selected instead; however, the author feels that the application to spectral theory given here is one of the most convincing and surprising.

In addition to basic functional analysis, the reader should possess some familiarity with measure theory (needed in Parts II and IV). Of non-elementary functional analytic theorems, only the bipolar theorem [15, IV.1.5], and the theorems of Krein-Milman [Kreĭn-Mil'man] [15, II.10.4] and Krein-Šmulian [15, IV.6.4] are used. With the exception of Part III and Part IV, §3, most of the results of this study are contained in the author's recent book [16], to which the interested reader is referred for a systematic exposition, references, historical notes, and many examples and applications.

PART I. RIESZ SPACES AND BANACH LATTICES

1. Definitions and Basic Algebraic Properties. The easiest and most intuitive way of defining a Riesz space (vector lattice) is through the concept of an ordered vector space. A vector space E over **R**, endowed with an order relation $\leqslant$ (i.e., a binary relation satisfying $x \leqslant x$, $x \leqslant y$ & $y \leqslant x \Rightarrow x = y$, and $x \leqslant y$ & $y \leqslant z \Rightarrow x \leqslant z$ for all $x,y,z \in E$), is called an *ordered vector space* if the ordering $\leqslant$ is invariant under translation and under multiplication by positive scalars:

(LO)$_1$ $\qquad x \leqslant y \Rightarrow x + z \leqslant y + z \qquad (x,y,z \in E),$

(LO)$_2$ $\qquad x \leqslant y \Rightarrow \lambda x \leqslant \lambda y \qquad (x,y \in E, \lambda \in \mathbf{R}_+).$

An ordered vector space $(E, \leqslant)$ is called a *Riesz space* (or a *vector lattice*) if any doubleton $\{x,y\} \subset E$ has a least upper bound $x \vee y := \sup\{x,y\}$ and a greatest lower bound $x \wedge y := \inf\{x,y\}$. The ordering of E is called *Archimedean* if $x \leqslant n^{-1}y$ for some pair (x,y) and all $n \in \mathbf{N}$, implies $x \leqslant 0$; in general, we will be concerned with Archimedean Riesz spaces.

Let E denote a Riesz space. In addition to the *lattice operations* $(x,y) \mapsto x \vee y$ and $(x,y) \mapsto x \wedge y$ from $E \times E$ into E, it has proved useful to define the mappings from E into E: $x \mapsto x^+ := x \vee 0$,

$x \mapsto x^- := (-x) \vee 0$, and $x \mapsto |x| := x \vee (-x)$. The elements x^+, x^-, and $|x|$ are called the *positive part*, the *negative part*, and the *modulus* of x, respectively; all are elements of the set $E_+ := \{x \in E: x \geq 0\}$ called the *positive cone* of E.

Certain relations are valid between the vector and lattice operations of E, of which we list the following $(x, y, x_1, y_1 \in E, \lambda \in \mathbf{R})$:

$$x + y = x \vee y + x \wedge y, \quad x = x^+ - x^-, \quad |x| = x^+ + x^-, \quad (1)$$

$$|x + y| \leq |x| + |y|, \quad |\lambda x| = |\lambda| |x|, \quad |x| = 0 \Leftrightarrow x = 0, \quad (2)$$

$$|x \vee y - x_1 \vee y_1| \leq |x - x_1| + |y - y_1|,$$

$$|x \wedge y - x_1 \wedge y_1| \leq |x - x_1| + |y - y_1|. \quad (3)$$

Proofs of these relations can be found in Birkhoff [3], Luxemburg-Zaanen [12], and Schaefer [16]. Rather than proving them here again, we point out that (as a consequence of certain representation theorems; cf. [16, Chapter II, Exercise 21]) a relation involving only a finite combination of $=$, $\leq$ and the vector and lattice operations (such as (1)—(3) above), is valid in every Archimedean vector lattice whenever it is valid in $\mathbf{R}$. An important consequence of (1) is the fact that the lattice structure of E is determined by the map $x \mapsto |x|$ alone; indeed, x and $|x|$ determine x^+, x^- and hence, by (LO)$_1$, $x \vee y = x + (y - x)^+$ and $x \wedge y = x - (y - x)^-$ for all $x, y \in E$.

We must mention another important concept: orthogonality. Elements x, y are called *orthogonal* (or *lattice disjoint*) if $|x| \wedge |y| = 0$; this is often denoted by $x \perp y$. A non-trivial characterization of orthogonality is the equivalence

$$x \perp y \Leftrightarrow |x + y| = |x - y|, \quad (4)$$

which is reminiscent of Hilbert space orthogonality; there are other analogies between the two concepts. On spaces where both notions are defined (essentially spaces L^2), lattice orthogonality is the far more stringent condition (cf. Part II, §2).

The simplest example, other than $\mathbf{R}$ itself, of a vector lattice is the n-dimensional number space $\mathbf{R}^n$ ($n \in \mathbf{N}$) under its canonical (coordinatewise) ordering defined by $|(\xi_1, \ldots, \xi_n)| := (|\xi_1|, \ldots, |\xi_n|)$.

We note in passing that any Archimedean vector lattice of finite linear dimension n is Riesz isomorphic to $\mathbf{R}^n$ (for the definition of a Riesz isomorphism, see below). More intriguing examples are the Banach sequence spaces c_0, c, and l^p ($1 \leq p \leq \infty$) of null, convergent, and p-summable sequences $x = (\xi_1, \xi_2, \ldots)$ with respective norms $\|x\|_\infty := \sup_n |\xi_n|$ (for c_0, c, and l^∞) and $\|x\|_p := (\sum_n |\xi_n|^p)^{1/p}$ ($1 \leq p < \infty$); l^∞ is the space of all bounded real sequences. The ordering is, of course, again defined coordinatewise. In the more general case of a measure space (X, Σ, μ) (see, e.g., [2]), the Banach spaces $L^p(\mu)$ ($1 \leq p \leq \infty$) become Riesz spaces if, for each equivalence class $[f]$, the modulus is defined to be the equivalence class $[|f|]$. A very important example is $C(K)$, the space of continuous functions on a compact space K.

In each of the preceding examples, the underlying space is a well-known Banach space. In addition, the following natural relation between the norm and modulus functions is fulfilled:

(NL) $\qquad\qquad |x| \leq |y| \Rightarrow \|x\| \leq \|y\|.$

A norm on a Riesz space satisfying (NL) is called a *Riesz norm* (or *lattice norm*); the triple $(E, \leq, \|\ \|)$ is called a *normed Riesz space* (or *normed vector lattice*). (Note that a normed Riesz space is necessarily Archimedean.) If, in addition, $(E, \|\ \|)$ is complete then we call E a *Banach lattice* (rather than "norm complete normed Riesz space"). Since (3) and (NL) imply that the lattice operations of a normed Riesz space E are uniformly continuous, they have unique continuous extensions to the norm completion $\tilde{E}$; the resulting triple $(\tilde{E}, \leq, \|\ \|)$ is a Banach lattice with positive cone $\tilde{E}_+$, $\tilde{E}_+$ denoting the closure of E_+ in $\tilde{E}$. Examples of Banach lattices other than those given above will appear in Sections 2, 3, and 5 below.

Another property of Riesz spaces, familiar from $\mathbf{R}$ and $\mathbf{R}^n$ ($n > 1$), must be mentioned explicitly. From the definitions it follows immediately that in a Riesz space E, each non-void finite subset A has a least upper bound $\sup A$ and a greatest lower bound $\inf A$. As the example of the Banach lattice c (convergent sequences) shows, in general this is no longer true for infinite subsets A of E, even if A is *order bounded* (i.e., contained in a suitable order interval $[x, y] := \{z \in E : x \leq z \leq y\}$ of E).

A Riesz space E is called *order complete* (or *Dedekind complete*) if each non-void order bounded subset $A \subset E$ possesses $\sup A$ and $\inf A$; the corresponding property for countable subsets is termed *order σ-completeness* (or *Dedekind σ-completeness*). The Banach lattices c_0, l^p and $L^p(\mu)$ ($1 \leq p < \infty$) are all order complete; $L^\infty(\mu)$ is order complete whenever μ is σ-finite. Every order σ-complete Riesz space is Archimedean.

2. Ideals and Projection Bands. In the preceding section we have remarked that any Archimedean Riesz space of finite dimension n is Riesz isomorphic to $\mathbf{R}^n$. What does this mean precisely? As in any theory of mathematical objects, an appropriate notion of "homomorphism" and "isomorphism" is indispensable for Riesz spaces. Given Riesz spaces E_1 and E_2, it is most natural to call a mapping $T: E_1 \to E_2$ a *Riesz homomorphism* if T preserves both the vector and lattice structures; that is, if T is linear and $|Tx| = T|x|$ for each $x \in E_1$. If, in addition, T is bijective, then T is called a *Riesz isomorphism*; E_1 and E_2 are called *Riesz isomorphic* if there exists a Riesz isomorphism between them. It is easy to see from (1) in §1 that each Riesz homomorphism T satisfies $T(x \vee y) = Tx \vee Ty$, $T(x \wedge y) = Tx \wedge Ty$, and in particular, $T(x^+) = (Tx)^+$, $T(x^-) = (Tx)^-$ for all elements x, y of its domain.

Let $T: E_1 \to E_2$ denote a Riesz homomorphism with range $F \subset E_2$ and kernel $J \subset E_1$. F is called a *Riesz subspace* (or *vector sublattice*) of E_2, and J is called an *ideal* (or *lattice ideal*) of E_1.[†] Ideals J of a Riesz space E are linear subspaces characterized by the property

(I) $\qquad (x \in J, y \in E \ \& \ |y| \leq |x|) \Rightarrow y \in J.$

In fact, it is tedious but not difficult to verify that if E is a Riesz space and J is a vector subspace satisfying (I), then E/J is a Riesz space under the ordering whose positive cone is the image of E_+ under the quotient map $E \to E/J$, and the latter map is a Riesz

---

[†]There are certain analogies between Riesz spaces and commutative algebras over $\mathbf{R}$, which extend well into the realm of Banach lattices and real commutative Banach algebras (cf. Part II).

homomorphism. (It should be noted that (I) does not imply that E/J is Archimedean, even if E is; for a detailed discussion, see [12, §60]. In our later applications, only ideals J for which E/J is Archimedean will be considered.)

To work efficiently with (I), we need the following basic result.

PROPOSITION 2.1 (Riesz Decomposition Property): *If E is a Riesz space and if $x \in E_+, y \in E_+$, then we have*

(D) $$[0,x] + [0,y] = [0,x+y].$$

Equivalently: *If $x,y \geqslant 0$ and if $0 \leqslant z \leqslant x+y$ in E, there exist elements x_1, y_1 of E satisfying $0 \leqslant x_1 \leqslant x$, $0 \leqslant y_1 \leqslant y$, and $z = x_1 + y_1$.*

Proof: Let $x_1 := x \wedge z$ and $y_1 := z - x_1$. Then we have $0 \leqslant x_1 \leqslant x, z = x_1 + y_1$, and $y_1 = z - z \wedge x = z + (-x) \vee (-z) = (z-x) \vee 0 \leqslant (x+y-x) \vee 0 = y$. □

It is very easy to see now that under set inclusion, the family $\mathbf{I}(E)$ of all ideals in E is a lattice. Let $I, J \in \mathbf{I}(E)$. It is clear from (I) that $I \cap J$ is an ideal equal to $\inf(I,J)$. On the other hand, $\sup(I,J)$ (=intersection of all ideals containing both I and J) must contain $I+J$; we show that $I+J$ is an ideal, whence $I+J = \sup(I,J)$. In fact let $u \in E$ satisfy $|u| \leqslant x+y$ for suitable elements $x \in I \cap E_+, y \in J \cap E_+$: then $u^+ \leqslant x+y$ and $u^- \leqslant x+y$ and from (D) it follows that $u^+ \in I+J$, $u^- \in I+J$. Since $I+J$ is a vector subspace of E, we have $u \in I+J$ as claimed.

The remarkable thing about ideals is the following result.

PROPOSITION 2.2: *For any Riesz space E, the lattice $\mathbf{I}(E)$ is distributive; that is, if $I, J, K \in \mathbf{I}(E)$, then*

$$(I+J) \cap K = I \cap K + J \cap K.$$

Proof: It is clear that $I \cap K + J \cap K \subset (I+J) \cap K$, so let us prove the converse relation. If $u \in (I+J) \cap K$, then $u = v + w$ where $v \in I$, $w \in J$; hence $|u| \leqslant |v| + |w|$ and, by (D), $|u| = x+y$ for suitable elements $x \in I_+, y \in J_+$. But since K is an ideal, we have

$|u| \in K$ and so $x \in K, y \in K$ which shows that $x \in I \cap K, y \in J \cap K$. Thus $|u| \in I \cap K + J \cap K$ and, since the latter set is an ideal as shown above, it follows that $u \in I \cap K + J \cap K$. □

Examples of ideals in the sequence spaces c_0, c, l^p are the sets $J_A := \{x = (\xi_n) : \xi_n = 0 \text{ for } n \in A\}$, A denoting any subset of $\mathbf{N}$. With the exception of c and l^∞, these ideals exhaust the supply. The reader will observe that if x_A denotes a sequence in any of these spaces whose nth coordinate is $\neq 0$ exactly when $n \in A$, then $J_A = \{x : |x| \wedge |x_A| = 0\}$. This leads naturally to the observation that for any Riesz space E and any subset A of E, $A^\perp := \{x \in E : |x| \wedge |y| = 0 \text{ for all } y \in A\}$ is an ideal of E; since $A^\perp$ is a linear subspace of E, verification is immediate from (I) above.

In particular, if $J \subset E$ is an ideal then $J^\perp$ is an ideal such that $J \cap J^\perp = \{0\}$, and no element $\neq 0$ of E is orthogonal to $J + J^\perp$. However, as may be seen from the example $E = C[0,1]$, $J = \{f \in E : f(0) = 0\}$, we cannot generally expect that $E = J + J^\perp$. Therefore, an ideal $J \subset E$ for which $E = J + J^\perp$ is called a *complemented ideal* or *projection band* of E. (In contrast with general ideals, bands have the additional property that they contain the suprema of arbitrary subsets, whenever these suprema exist in E.) As with any direct sum decomposition of a vector space, with a band decomposition $E = J + J^\perp$ there is associated a projection $P: E \to J$ with range J and kernel $J^\perp$. It is easy to show (see the proof of Proposition 2.4 below) that, in addition to $P^2 = P$, these *band projections* are characterized by the property that $0 \leq Px \leq x$ for all $x \in E_+$.

PROPOSITION 2.3: *Let E be any Riesz space. Under set inclusion, the family $\mathbf{B}(E)$ of all projection bands of E is a sublattice of $\mathbf{I}(E)$, and a Boolean algebra.*

Proof: If $E = I + I^\perp = J + J^\perp$ for $I, J \in \mathbf{B}(E)$, distributivity (Proposition 2.2) shows that
$$E = (I + I^\perp) \cap (J + J^\perp) = (I \cap J) + [I \cap J^\perp + I^\perp \cap J + I^\perp \cap J^\perp],$$
and the second summand is found to be $(I \cap J)^\perp$; therefore $I \cap J \in \mathbf{B}(E)$ and, similarly, $I + J \in \mathbf{B}(E)$. □

Let us note that in general, $\mathbf{B}(E)$ may not contain projection bands other than $\{0\}$ and E; for example, such is the case for $E = C[0, 1]$. However, if E is order complete, then for each subset $A \subset E$, $E = B_A + A^\perp$ where B_A denotes the smallest band (i.e., the smallest ideal closed under the formation of arbitrary suprema in E) containing A. (Riesz decomposition theorem; for a proof see, e.g., [16, I.2.10].)

Now let E denote a normed Riesz space. If F is a Riesz subspace of E (i.e., a vector subspace such that $x \in F$ implies $|x| \in F$), then its closure $\bar{F}$ is a Riesz subspace; this follows at once from the continuity of the lattice operations (cf. (3) in §1). Similarly, the closure of an ideal J is an ideal $\bar{J}$. In fact if $y \in \bar{J}$, $z \in E$ and $|z| \leq |y|$, then $|y| \in \bar{J}$ by the preceding. Now if (x_n) is a sequence in J_+ with limit $|y|$, then $x_n \wedge |z| \in J$ and $\lim_n x_n \wedge |z| = |y| \wedge |z| = |z|$; so $|z| \in \bar{J}$. Likewise we have $z^+ \in \bar{J}$ and $z^- \in \bar{J}$, whence $z \in \bar{J}$.

PROPOSITION 2.4: *If I is a projection band in a normed Riesz space E, the associated band projection $P: E \to I$ has norm ≤ 1. In particular, each projection band in E is closed.*

Proof: We note first that if $E = I + I^\perp$ and $x = x_1 + x_2$ is the corresponding decomposition of $x \in E$, then $x \geq 0$ implies that $x_1 \geq 0$ and $x_2 \geq 0$. In fact $0 \leq x_1^+ - x_1^- + x_2^+ - x_2^-$ implies that $0 \leq x_1^- + x_2^- \leq x_1^+ + x_2^+$, and $x_1^- + x_2^- \perp x_1^+ + x_2^+$ shows that $x_1^- = 0$, $x_2^- = 0$. It follows that for $x = x_1 + x_2$ where $x_1 \in I$, $x_2 \in I^\perp$ but x is not necessarily positive, we must have $|x| = |x_1| + |x_2|$. Therefore $\|x_1\| = \| |x_1| \| \leq \| |x| \| = \|x\|$ and the projection $x \mapsto x_1$ has norm ≤ 1; for reasons of symmetry, $x \mapsto x_2$ has norm ≤ 1. Finally $I = PE$ is closed, since $I = (1_E - P)^{-1}(0)$. □

We have remarked above that for any Riesz space E and ideal J of E, E/J is a Riesz space; if E is a normed Riesz space and J closed, the quotient norm on E/J is a Riesz norm [16, II.5.4] (in particular, the quotient of a Banach lattice over a closed ideal is a Banach lattice). Maximal ideals do not always exist in Banach lattices, but it can be shown that a maximal ideal in a Banach lattice is necessarily closed (see §5 below).

3. Duality of Riesz Spaces. As in general functional analysis, the study of appropriate classes of linear forms (linear functionals) provides the key to many fruitful insights; illustrative examples will be considered below (§§4, 5). A linear form $f: E \to \mathbf{R}$ on a Riesz space E is called *order bounded* if for each order interval $[x,y]$ of E, $f([x,y])$ is a bounded subset of $\mathbf{R}$. It is clear that the set E^* of all order bounded linear forms is a vector subspace of the algebraic dual of E; E^* is called the *order dual* of E.

PROPOSITION 3.1: *Let E be any Riesz space. Under the ordering defined by "$f \geq 0$ iff $f(x) \geq 0$ for all $x \in E_+$," E^* is an order complete Riesz space.*

Proof: Let $f \in E^*$ be given and define, for each $x \in E_+$, the number

$$r(x) := \sup\{f(y) - f(z): y, z \in E_+, y + z = x\}.$$

Clearly $r(\lambda x) = \lambda r(x)$ whenever $\lambda \geq 0$; moreover, since $[0, x_1 + x_2] = [0, x_1] + [0, x_2]$ by the decomposition property (D) (cf. 2.1), it is readily seen that $r(x_1 + x_2) = r(x_1) + r(x_2)$ for all $x_1, x_2 \in E_+$. Thus $r: E_+ \to \mathbf{R}_+$ is additive and positive homogeneous; in addition, if $s: E_+ \to \mathbf{R}_+$ is a mapping with these two properties and satisfying $s(x) \geq f(x) \vee -f(x)$ for all $x \in E_+$, then $s \geq r$.

Now if $z \in E$ and $z = x - y$ is any decomposition with $x, y \in E_+$, the number $h(z) := r(x) - r(y)$ is independent of the chosen decomposition of z (in particular, $h(z) = r(z^+) - r(z^-)$), and so $z \mapsto h(z)$ is a linear form on E. By its construction and the ordering of E^* defined above, we have $h \geq f$, $h \geq -f$, and $h \leq g$ for any linear form g majorizing both f and $-f$; therefore, $h = f \vee (-f) = |f|$ and E^* is seen to be a Riesz space. The order completeness of E^* is almost immediate: If $A \subset E^*$ is a directed ($\leq$), order bounded set (we can assume that $A \subset [0, h]$, say) then

$$t(x) := \sup_{f \in A} f(x) \qquad (x \in E_+)$$

defines an additive, positive homogeneous map $t: E_+ \to \mathbf{R}_+$ which extends to a positive linear form g as above; clearly we have $g = \sup A$ in E^*. □

In the preceding proof, the modulus $|f|$ of $f \in E^*$ was defined at $x \in E_+$ by $|f|(x) = \sup\{f(y) - f(z): y, z \in E_+, y + z = x\}$; similarly, the following formulas are obtained for $f, g \in E^*$.

$$f \vee g(x) = \sup\{f(y) + g(z): y, z \in E_+, y + z = x\},$$
$$f \wedge g(x) = \inf\{f(y) + g(z): y, z \in E_+, y + z = x\}, \quad (1)$$
$$|f|(x) = \sup\{|f(u)|: u \in E, |u| \leq x\},$$

are all valid for any $x \in E_+$. An order bounded linear form f on E is called *order continuous* if for any directed ($\geq$) subset A of E, $\inf A = 0$ implies $\inf |f(A)| = 0$. We mention without proof that the set E^*_{oc} of all order continuous linear forms is an ideal (even a projection band) of E^*; for details, see [11] or [16, II.4].

The positive elements of E^*, E any Riesz space, are called *positive linear forms* on E. Let us note that a positive linear form $f \neq 0$ on E is a Riesz homomorphism $E \to \mathbf{R}$ if and only if its kernel $f^{-1}(0)$ is a maximal ideal of E. In fact, if $|f(x)| = f(|x|)$ for all $x \in E$ then $f^{-1}(0)$ is an ideal of E and of codimension 1, hence maximal; conversely, if $f > 0$ and $J := f^{-1}(0)$ is an ideal of E, then f induces a Riesz isomorphism of E/J onto $\mathbf{R}$. Also, it is not difficult to verify that a non-zero Riesz homomorphism $f: E \to \mathbf{R}$ generates an extreme ray of E^*_+, and conversely (cf. II.1.1 below).

Later on we will be concerned with the following problem: If E is a Riesz space and if G is a Riesz subspace of E^*, how does E imbed into G^* by evaluation? In other words, under what conditions on G is the mapping $q: E \to G^*$ which to each $x \in E$ orders the linear form $f \mapsto f(x)$ on G, necessarily a Riesz homomorphism? The answer is affirmative whenever G is an ideal of E^*.

PROPOSITION 3.2: *If E is a Riesz space and if G is an ideal of E^*, then the evaluation map $q: E \to G^*$ is a Riesz homomorphism.*

Proof: For each $x \in E$, let us denote by $\tilde{x}$ the element $q(x)$. Obviously q is linear, and we have to show that $(\tilde{x})^+ = (x^+)^\sim$. Now $x \leq x^+$ implies $f(x) \leq f(x^+)$ for all $f \in G_+$, hence $\tilde{x} \leq (x^+)^\sim$ and so $(\tilde{x})^+ \leq (x^+)^\sim$. To prove the converse inequality, for each $f \in G_+$

we construct an element $g_f \in G_+$ by defining

$$g_f(y) := \sup\{f(z): z \in [0,y] \cap P\} \quad (y \in E_+),$$

where $P := \bigcup_{n=1}^{\infty} n[0, x^+]$. From Proposition 2.1 it follows that g_f is additive and positive homogeneous on E_+; its linear extension to E (which we again denote by g_f) satisfies $0 \leq g_f \leq f$ in E^*. Using the ideal property of G, we conclude that $g_f \in G_+$; clearly we have $g_f(x^-) = 0$. Thus

$$(x^+)^{\sim}(f) = f(x^+) = g_f(x^+) = g_f(x) \leq \sup_{0 \leq h \leq f} h(x) = (\tilde{x})^+(f)$$

by definition of $(\tilde{x})^+$ (see (1) above). Since $f \in G_+$ was arbitrary, we obtain the desired inequality $(x^+)^{\sim} \leq (\tilde{x})^+$. □

We must now look briefly at the order properties of the Banach dual of a normed Riesz space; we recall that the dual E' of a normed space $(E, \| \ \|)$ is a Banach space under the dual norm $\|f\| := \sup\{|f(x)|: \|x\| \leq 1\}$.

PROPOSITION 3.3: *Let E denote a normed Riesz space. The topological dual E' is an ideal of E^*, and E' is an order complete Banach lattice under the dual norm.*

Proof: Let U denote the closed unit ball of E. Since $x \in U$ implies $|x| \in U$ by Axiom (NL) (§1), for each positive $f \in E'$ we have

$$\|f\| = \sup_{\|x\| \leq 1} f(|x|) = \sup_{\|x\| \leq 1, x \geq 0} f(x). \quad (2)$$

Now let $g \in E'$ satisfy $\|g\| \leq 1$. By Formula (1) we have $|g|(|x|) \leq 1$ for all $x \in U$, and hence $|g| \in E'$ and $\| |g| \| \leq 1$. By (2) above, $|g| \leq |f|$ implies that $\|g\| \leq \|f\|$ so Axiom (NL) is satisfied. Finally, if $f \in E'$, $g \in E^*$, and if $|g| \leq |f|$, then $|g(x)| \leq |g|(|x|) \leq |f|(|x|) \leq \|f\| \|x\|$, which implies that $g \in E'$. Therefore, E' is an ideal of E^* and it follows from Proposition 3.1 that E' is order complete. □

COROLLARY: *If E'' denotes the ordered Banach bidual of a normed Riesz space E, then E'' is an order complete Banach lattice and evaluation $E \to E''$ is a Riesz homomorphism.*

The proof of this corollary results at once from Propositions 3.2 and 3.3; in addition, we note that the norm of $x \in E_+$ is given by

$$\|x\| = \sup\{\langle x, x' \rangle : x' \in E'_+, \|x'\| \leq 1\}. \tag{3}$$

We conclude this section with a duality theorem on normed Riesz spaces which will be basic for the extension theorems of Part III.

THEOREM 3.4: *Let E, F be normed Riesz spaces with respective closed unit balls U, V, and let $U^\circ_+ := E'_+ \cap U^\circ$ and $V^\circ_+ := F'_+ \cap V^\circ$ denote the positive parts of the dual unit balls. If $j: F \to E$ is an isometric Riesz homomorphism, then its adjoint $j': E' \to F'$ maps U°_+ onto V°_+.*

Proof: If under j, the space F is identified with a normed Riesz subspace of E, our claim amounts to the assertion that every positive linear form f of norm ≤ 1 on F has a norm-preserving, positive linear extension to E. Observing that the hypothesis on f is equivalent to the assertion that $f(x) \leq \|x^+\|$ for all $x \in F$, it suffices for the proof to apply the classical Hahn-Banach theorem to the sublinear functional $x \mapsto \|x^+\|$ on E. □

4. Relations Between Norm and Order. The relationship between norm and order implied by Axiom (NL) (§1) is weaker than might be expected. For example, if $E = C[0,1]$, if $Q \neq \emptyset$ is a nowhere dense subset of $[0,1]$, and if A is the set of all functions f in E satisfying $0 \leq f \leq e$ (e the constant-one function) and $f(Q) = \{1\}$; then A is a net (under its natural direction $\geq$) such that $\inf A = 0$ and $\inf\{\|f\|: f \in A\} = 1$. That is, A order converges to 0 but is far from converging to 0 in norm. By contrast, if (f_n) is a decreasing sequence in $L^1[0,1]$ (with Lebesgue measure) converging a.e. to a function $f \in L^1$, then $\lim_n \|f_n - f\| = 0$ by Lebesgue's dominated convergence theorem, and a corresponding result holds for arbitrary directed sets in L^1 (see 4.4 below).

A Banach lattice E is said to have *order continuous norm* if for each directed ($\geqslant$) subset A of E, $\inf A = 0$ implies $\lim A = 0$. Examples of Banach lattices with order continuous norm are the spaces c_0, l^p, $L^p(\mu)$ ($1 \leqslant p < \infty$) considered above (§1); the norms of c, l^∞, $L^\infty(\mu)$ are not order continuous. Our main objective in this section is to characterize Banach lattices with order continuous norm in various ways.

PROPOSITION 4.1: *Let E be a normed Riesz space. If $A \subset E_+$ is a non-void directed ($\geqslant$) set such that $\lim_A f(x) = 0$ for each $f \in E'_+$, then $\lim_A \|x\| = 0$.*

Proof: By means of evaluation, each $x \in E_+$ can be considered to be a weak* continuous function on the set $U^\circ_+ = \{f \in E'_+ : \|f\| \leqslant 1\}$; the hypothesis then states that A converges to 0 pointwise on U°_+. Since A is directed ($\geqslant$) and U°_+ is weak* compact, the classical Dini theorem shows the convergence to be uniform and the assertion follows from §3, Formula (3). □

PROPOSITION 4.2: *Let E be a normed Riesz space, and let A be a directed ($\leqslant$) subset which converges (as a net) to $x \in E$. Then $x = \sup A$.*

Proof: Since $E_+ = \{x \in E : x^- = 0\}$ and $x \mapsto x^-$ is continuous, the positive cone E_+ is closed in E. Thus for each $z \in A$ we have $x \in z + E_+$ or $x \geqslant z$. On the other hand, if y is an upper bound of A then $A \subset y - E_+$; since $y - E_+$ is closed, we have $x \in y - E_+$ or $x \leqslant y$. □

For the announced characterization of Banach lattices with order continuous norm, we also need the following lemma.

LEMMA 4.3 (Dixmier): *Let K be a compact space such that $C(K)$ is order complete. If N is a set of order continuous linear forms that separates $C(K)$, then the closed unit ball of $C(K)$ is compact for the weak topology $\sigma(C(K), N)$.*

A proof will be given in Part II below (II.3.3, Corollary 1). In addition, we must borrow from Part II the following result: If E is a Banach lattice and if $x \in E_+$, then the ideal given by $E_x := \bigcup_{n=1}^{\infty} n[-x,x]$, normed by the gauge of $[-x,x]$, is Riesz and norm isomorphic to $C(K)$ for a suitable compact space K.

THEOREM 4.4: *The following properties of a Banach lattice E are equivalent*:

(a) *E has order continuous norm.*

(b) *Every continuous linear form on E is order continuous.*

(c) *Evaluation $E \to E''$ maps E onto an ideal of E''.*

(d) *Every order interval in E is weakly compact.*

(e) *E is order σ-complete, and no Banach sublattice is Riesz isomorphic to l^∞.*

(f) *Every order bounded, orthogonal sequence in E converges to 0.*

Proof: (a)$\Rightarrow$(b) is clear from the definitions.

(b)$\Rightarrow$(d): It suffices to consider order intervals of the form $[-x,x]$ ($x \in E_+$), since each order interval $[x,y]$ is contained (and closed) in the symmetric interval $[-|x|-|y|,|x|+|y|]$. By the remark preceding (4.4), we have $E_x \cong C(K)$ and by (b), the restrictions to E_x of the elements of E' form a separating set N of order continuous linear forms on $C(K)$; so by Lemma 4.3, $[-x,x]$ is $\sigma(E,E')$-compact.

(c)$\Leftrightarrow$(d): Suppose E is an ideal of E'' under evaluation. Identifying E with its image, we obtain $[x,y] = (x + E''_+) \cap (y - E''_+)$ for any order interval $[x,y]$ of E, and hence $[x,y]$ is $\sigma(E'',E')$-closed in E''. Since it is also bounded, $[x,y]$ is $\sigma(E'',E')$-compact and thus (as a subset of E) $\sigma(E,E')$-compact. Conversely, suppose each order interval in E to be weakly compact, and let $-x \leqslant z \leqslant x$ for some $x \in E$ and $z \in E''$. Then we have $z \in (-x + E''_+) \cap (x - E''_+)$; but the latter set is the bipolar of $[-x,x] \subset E$ with respect to the duality $\langle E'', E' \rangle$ and so, by the weak compactness of $[-x,x]$ and the bipolar theorem, we obtain $(-x + E''_+) \cap (x - E''_+) = [-x,x] \subset E$. Thus $z \in E$ and it follows that E is an ideal of E''.

(d)$\Rightarrow$(e): Let $A \subset [x,y]$ denote a directed ($\leqslant$) order bounded subset of E. By the weak compactness of $[x,y]$, the net A has a weak limit $z \in [x,y]$ which, since E_+ is weakly closed, implies that $z = \sup A$. (See the proof of Proposition 4.2 and note that by

Proposition 4.1, $z = \lim A$ in norm which leads back to (a).) Moreover, if l^∞ were Riesz isomorphic to a Banach sublattice of E, (d) would imply the closed unit ball of l^∞ to be weakly compact, which is false.

(e)$\Rightarrow$(f): Suppose (x_n) is an (infinite) order bounded, orthogonal sequence in E which does not converge to 0; we can assume that $x_n \in [0,x]$ and $\|x_n\| = 1$ for all n. We define a mapping $\phi: l_+^\infty \to E_+$ by letting $\phi(\alpha_1, \alpha_2, \ldots) := \sum_n \alpha_n x_n := \sup_n (\sum_{\nu=1}^n \alpha_\nu x_\nu)$; the supremum exists, since E is order σ-complete and since $\sum_{\nu=1}^n \alpha_\nu x_\nu \leqslant (\sup_n \alpha_n) x$ for all n. Clearly, ϕ extends to a Riesz homomorphism of l^∞ into E; taking norms in the last inequality we obtain $|\alpha_n| = \|\alpha_n x_n\| \leqslant \|\sum_n \alpha_n x_n\| = \|\sum_n |\alpha_n| x_n\| \leqslant (\sup_n |\alpha_n|)\|x\|$. Therefore, $\sup_n |\alpha_n| \leqslant \|\sum_n \alpha_n x_n\| \leqslant (\sup_n |\alpha_n|)\|x\|$ and this shows ϕ to be a Riesz isomorphism of l^∞ onto a Banach sublattice of E, contrary to the hypothesis.

(f)$\Rightarrow$(a): Let A be a directed ($\geqslant$) family in $[0,x] \subset E_+$ such that $\inf A = 0$; we have to show that $\lim A = 0$. As above we identify the ideal E_x with a space $C(K)$ (K compact); then x becomes the constant-one function e and we have $A \subset [0,e]$. For a given real $\epsilon > 0$ define $Q_\epsilon := \{t \in K: f(t) \geqslant \epsilon, \text{ all } f \in A\}$; because $\inf A = 0$, the set Q_ϵ is a nowhere dense subset of K. Now let B_ϵ denote the set of all $g \in [0,e]$ such that $U_g := \{t \in K: g(t) = 1\}$ is a neighborhood of Q_ϵ. Under the relation $\prec$, defined by "$g_2 \prec g_1$ if $g_2 = g_1$ or if $g_2(s) = 0$ whenever $s \notin U_{g_1}$," B_ϵ is a net. Moreover, since the directed ($\geqslant$) family $\{(f - \epsilon e)^+: f \in A\}$ converges to 0 pointwise (hence uniformly) in the complement of each neighborhood of Q_ϵ, for a given $g \in B_\epsilon$ there exists $f \in A$ satisfying $(f - \epsilon e)^+ \leqslant g + \epsilon e$ and hence, satisfying $f \leqslant g + 2\epsilon e$. Consequently, if we have $\lim B_\epsilon = 0$ (in E) for each $\epsilon > 0$, then we also have $\lim A = 0$. Otherwise, there exists an $\epsilon > 0$ such that the net B_ϵ is not Cauchy in E and hence, there exists a sequence $g_1 \succ g_2 \succ \cdots$ in B_ϵ such that $\|g_{n+1} - g_n\| > \delta$ for all n and some $\delta > 0$. But $|g_{n+1} - g_n|$ is an orthogonal sequence contained in $[0,e]$, and this contradicts (f). $\square$

We note from part (d)$\Rightarrow$(e) of the preceding proof that a Banach lattice with order continuous norm is necessarily order complete. Also, from (e) of Theorem 4.4 we obtain the following corollary.

COROLLARY: *Every separable, order σ-complete Banach lattice has order continuous norm.*

We conclude this section with a property that will be needed in Part IV below (and which, in fact, characterizes Banach lattices with order continuous norm [16, III.11.1]).

PROPOSITION 4.5: *If E has order continuous norm, then each closed ideal of E is a projection band.*

Proof: Let I denote a closed ideal of E. For each $x \in E_+$ we define $Px := \sup[0,x] \cap I$; because E has order continuous norm, the supremum exists and is contained in I. From the decomposition property (D) of Proposition 2.1 it follows that $P(x+y) = Px + Py$ whenever $x,y \geq 0$ and, clearly, we have $P(\lambda x) = \lambda Px$ if $\lambda \geq 0$; moreover, $P^2 x = Px$ and $0 \leq Px \leq x$, and of course $Px = x$ for $x \in I_+$. Defining $Pz := Pz^+ - Pz^-$ extends P to a projection of E onto I; since $Pz^+ \wedge Pz^- \leq z^+ \wedge z^- = 0$, Pz^+ must equal $(Pz)^+$ and so P is a Riesz homomorphism. This implies that $J := P^{-1}(0)$ is an ideal of E. Since $E = I + J$ and $I \cap J = \{0\}$, we must have $J = I^\perp$ and I is a projection band as claimed. □

5. Positive and Order Bounded Operators. Let E, F denote Riesz spaces. A linear operator $T: E \to F$ is called *positive* (or *increasing*, or *isotone*) if $x \geq 0$ implies $Tx \geq 0$, or equivalently, if $T(E_+) \subset F_+$; examples are the Riesz homomorphisms and the positive linear forms considered above. We begin with the following simple but important result.

THEOREM 5.1: *Let E be a Banach lattice and let F be a normed Riesz space. Every positive (linear) operator $T: E \to F$ is continuous.*

Proof: Suppose that $T: E \to F$ is positive and unbounded; then T cannot be bounded on the positive part of the unit ball of E. Hence, there exists a normalized sequence (x_n) in E_+ such that $\|Tx_n\| \geq n^3$ for all n. The series $\sum_n n^{-2} x_n$ converges to some $z \in E_+$ (since E_+ is norm complete) and we have $z \geq n^{-2} x_n$ for each n (cf. 4.2). Because T is positive, we obtain $Tz \geq n^{-2} Tx_n$ which implies that $\|Tz\| \geq n^{-2} \|Tx_n\| \geq n$, a contradiction. □

COROLLARY 1: *For every Banach lattice E, the topological dual E' agrees with the order dual E^*.*

This follows at once from Proposition 3.1 and the continuity of each positive linear form on E.

COROLLARY 2: *Each maximal ideal of a Banach lattice is closed.*

Proof: We have observed above (in the discussion preceding Proposition 3.2) that each maximal ideal of E is the kernel $f^{-1}(0)$ of a Riesz homomorphism $f: E \to \mathbf{R}$. Since f is positive and hence continuous, $f^{-1}(0)$ is closed. $\square$

COROLLARY 3: *If a Riesz space E is a Banach lattice under two distinct norms, the norms are equivalent.*

Proof: Denote by E_1 and E_2 two Banach lattices having the same underlying Riesz space E. Since the identity map of E defines positive (hence, continuous) bijections $E_1 \to E_2$ and $E_2 \to E_1$, it follows at once that the respective norms are equivalent. $\square$

Again, let E, F denote Riesz spaces. A linear map $T: E \to F$ is called *order bounded* if for each pair $(x,y) \in E \times E$, there exists a pair $(u,v) \in F \times F$ such that $T([x,y]) \subset [u,v]$. Every positive operator $T: E \to F$ is order bounded because $T([x,y]) \subset [Tx, Ty]$; obviously, each difference (and, more generally, each real linear combination) of positive operators is order bounded. The converse is not true in general, but it holds in the following important case.

PROPOSITION 5.2: *Let E, F be Riesz spaces and suppose F is order complete. The vector space $L^r(E,F)$ of all order bounded linear maps $E \to F$ is spanned by the set of all positive operators $E \to F$. Under the ordering defined by "$T_1 \leq T_2$ iff $T_1 x \leq T_2 x$ for all $x \in E_+$," $L^r(E,F)$ is an order complete Riesz space.*

Proof: Assuming F to be order complete, the proof that $L^r(E,F)$ is an order complete Riesz space under the ordering indicated is entirely analogous to the proof of Proposition 3.1. The

formulas
$$T \vee S(x) = \sup\{Ty + Sz : y, z \in E_+, y + z = x\},$$
$$T \wedge S(x) = \inf\{Ty + Sz : y, z \in E_+, y + z = x\}, \quad (1)$$
$$|T|(x) = \sup\{|Tu| : u \in E, |u| \leq x\},$$

which are valid for $x \in E_+$, are established in the same way except that the right-hand suprema and infima are to be taken in F (it is here that the assumption on F enters decisively). The mappings $E_+ \to F$ defined in (1) are then extended to E in the unique manner consistent with linearity. Similarly, if (T_α) is a directed ($\leq$) family of order bounded operators majorized by some $S \in L^r(E, F)$, the formula $Tx := \sup_\alpha T_\alpha x$ ($x \in E_+$) defines $T = \sup_\alpha T_\alpha$. □

Traditionally, differences of positive operators have been called *regular*; this accounts for the superscript "r". It is now also clear that using the notation $T \geq 0$ for a positive operator T, is perfectly legitimate; F being supposed order complete, "$\leq$" denotes in fact a Riesz space ordering of $L^r(E, F)$. Finally, we observe that the tensor product $E^* \otimes F$ can be considered as a linear subspace of $L^r(E, F)$; in fact, if as usual $u = \sum x_i^* \otimes y_i$ is interpreted as the linear map $x \mapsto u(x) = \sum \langle x, x_i^* \rangle y_i$, then $\pm u \leq \sum |x_i^*| \otimes |y_i|$ and so u is order bounded. (However, the modulus $|u| \in L^r(E, F)$ is in general not an element of $E^* \otimes F$; cf. Part III. §2.)

We conclude these simple considerations by investigating the case where E and F are Banach lattices, F being supposed order complete. If $T: E \to F$ is positive then we have, in complete analogy to §3, (2), the formula

$$\|T\| = \sup\{\|Tx\| : x \geq 0, \|x\| \leq 1\}. \quad (2)$$

For an operator $T \in L^r(E, F)$, we define the *r-norm* by

$$\|T\|_r := \sup\{\||T|x\| : x \geq 0, \|x\| \leq 1\}. \quad (3)$$

Thus the r-norm of T is nothing but the operator norm of $|T|$ (note that, by Theorem 5.1, $|T|$ is continuous). Since $|T_1 + T_2| \leq |T_1| + |T_2|$, it is clear that (3) defines a norm on $L^r(E, F)$.

PROPOSITION 5.3: *Let E, F be Banach lattices and suppose F order complete. Under the norm (3), $L^r(E, F)$ is an order complete Banach lattice which will be denoted by $\mathcal{L}^r(E, F)$.*

Proof: It is obvious from (2) and (3) above that the r-norm is a Riesz norm (Axiom (NL), §1); so we have only to show that $\mathcal{L}^r(E, F)$ is norm complete. Let (T_n) be a Cauchy sequence in $\mathcal{L}^r(E, F)$; by selecting a subsequence, if necessary, we may assume that $\|T_n - T_{n+1}\|_r < 2^{-n}$ for all n. Because (by (3)) the r-norm is greater than the operator norm, (T_n) converges in the operator norm to a bounded linear operator $T: E \to F$. This operator T is at once seen to satisfy

$$|T - T_n|x \leqslant \sum_{\nu=n+1}^{\infty} |T_{\nu+1} - T_\nu|x$$

for each $x \in E_+$. This implies that T is order bounded and that $\||T - T_n|x\| \leqslant 2^{-n}$ uniformly for $x \in E_+$, $\|x\| \leqslant 1$; thus we obtain $\|T - T_n\|_r \leqslant 2^{-n}$ and the proof is complete. □

In general, $\mathcal{L}^r(E, F)$ is properly contained in $\mathcal{L}(E, F)$. One of two notable exceptions occurs when F is an order complete space $C(K)$.

PROPOSITION 5.4: *If E is any Banach lattice and if F is an order complete $C(K)$ (K compact), then $\mathcal{L}^r(E, F) = \mathcal{L}(E, F)$ with identity of the respective norms.*

Proof: First, since the closed unit ball of $C(K)$ is the order interval $[-e, e]$ (e the constant-one function), it is clear that each $T \in \mathcal{L}(E, F))$ is order bounded. Second, for any non-void order bounded subset A of $C(K)_+$ we have $\|\sup A\| = \sup\{\|f\| : f \in A\}$, and so we have $\|\sup\{|Tu| : |u| \leqslant x\}\| = \sup\{\|Tu\| : |u| \leqslant x\}$ for each $T \in \mathcal{L}(E, F)$ and $x \in E_+$. Therefore, $\|T\| = \| |T| \| = \|T\|_r$ as asserted. □

The other important case yielding identity between $\mathcal{L}^r(E, F)$ and $\mathcal{L}(E, F)$ occurs when E is a space $L^1(\mu)$ and when F is a Banach lattice which is the range of a positive, contractive projection in its bidual F''. For details and further related results, we must refer the interested reader to [16, Chapter 4, §1 and Exercise 16].

PART II. RELATIONS WITH FUNCTION SPACES

1. Continuous Functions on Compacta. It will be amply evident from the following that in Banach lattice theory, the spaces of continuous functions $C(K)$ (K compact) are all-important; for a recent encyclopedic monograph on this subject, we refer to Semadeni [17]. The present section centers on the well-known representation theorem (1.3), independently due to Kakutani [1941] and M. and S. Krein [1940], as well as some related facts needed later.

We begin with a geometric characterization of real-valued Riesz homomorphisms.

PROPOSITION 1.1: *For a linear form $f > 0$ on a Riesz space E, the following assertions are equivalent*:
(a) f *is a Riesz homomorphism*.
(b) f *generates an extreme ray of the dual cone E_+^**.

Proof: (a)⇒(b): If f is a Riesz homomorphism, its kernel $N := f^{-1}(0)$ is an ideal of codimension 1 (hence maximal). Its annihilator N° in the algebraic dual of E is thus of dimension 1 and is easily seen to be an ideal of E^*. Therefore, if $g \in E^*$ and $|g| \leq f$ then $g = \lambda f$ for some $\lambda \in \mathbf{R}$. Thus $\{\lambda f \colon \lambda \geq 0\}$ is an extreme ray of E_+^*.

(b)⇒(a): If f generates an extreme ray of E_+^*, then the linear hull of $\{f\}$ is a one-dimensional ideal G of E^*, and the annihilator G° in E is the kernel of the evaluation map $q \colon E \to G^*$. Since q is a Riesz homomorphism by Proposition I.3.2 and since G° is also the kernel of f, then f is a Riesz homomorphism. □

We also need the following well-known version of the Stone-Weierstrass theorem, whose proof can be found in most functional analytic texts (for example, in [16, II.7.3]).

LEMMA 1.2: *Let K be compact. If $F \subset C(K)$ is a Riesz subspace containing the constant-one function e and separating the points of K, then F is dense in $C(K)$ (for the supremum norm).*

A Banach lattice E is called an *abstract* M-*space* (or an AM-*space*) if its norm satisfies

(M) $\qquad \|x \vee y\| = \|x\| \vee \|y\| \quad$ for all $x, y \in E_+$.

This condition is, in particular, satisfied if the closed unit ball U of E contains a greatest element e (so that $U = [-e, e]$); these Banach lattices are called AM-*spaces with unit*. While c_0 and $C_0(X)$ (= continuous functions on a locally compact space X vanishing at infinity) are AM-spaces without unit, examples of AM-spaces with unit are furnished by c, $L^\infty(\mu)$, and $C(K)$.

While a great variety of AM-spaces without unit has been discovered recently (Goullet de Rugy [9]), the announced representation theorem shows that every AM-space with unit is (Riesz and norm) isomorphic to $C(K)$ for a unique compact space K (see the Corollary below).

THEOREM 1.3 (Kakutani-Krein): *Every* AM-*space E with unit e is isomorphic to some $C(K)$. More precisely: If K denotes the weak* compact extreme boundary of $\{f \in E'_+ : f(e) = 1\}$, then evaluation at the points of K defines a Riesz and norm isomorphism of E onto $C(K)$.*

Proof: Since the theorem and its proof are very well known, we can be brief. Clearly, $H := \{f \in E'_+ : f(e) = 1\}$ is weak* compact in E'; since each extreme ray of E'_+ is generated by exactly one extreme point of H, Proposition 1.1 implies that the extreme boundary K of H agrees with the set of all Riesz homomorphisms $f : E \to \mathbf{R}$ normalized to satisfy $f(e) \, (= \|f\|) = 1$. This latter characterization shows K to be weak* closed, hence compact. Thus the map $q : x \mapsto g_x$ (where $g_x(f) := f(x)$ for all $f \in K$) is a Riesz homomorphism, and by Part I, §3, (3), q is an isometry of E into $C(K)$. Since E separates the points of K and $q(e)$ is the constant-one function on K, it follows from Lemma 1.2 that $q(E)$ is dense in $C(K)$. But E is norm complete and so $q(E) = C(K)$. □

COROLLARY: *Every Riesz homomorphism $C(K) \to \mathbf{R}$ is of the form $g \to \alpha g(t_0)$ for suitable $\alpha \geq 0$ and $t_0 \in K$, and conversely. Thus the normalized real Riesz homomorphisms of $C(K)$ correspond bijectively to the points of K. In particular, $C(K_1)$ and $C(K_2)$ (K_1, K_2 compact) are Riesz isomorphic if and only if K_1 and K_2 are homeomorphic.*

One reason for the importance of the spaces $C(K)$ is their occurrence as principal ideals of arbitrary Banach lattices. In fact, if E is a Banach lattice and if $u \in E_+$, the ideal of E generated by $\{u\}$ consists of all $x \in E$ satisfying $|x| \leq cu$ for suitable $c \in \mathbf{R}_+$; that is, $E_u = \bigcup_{n=1}^{\infty} n[-u, u]$. Since $[-u, u]$ is a convex, circled, absorbing subset of E_u containing no vector subspace other than $\{0\}$, its gauge function p is a norm on E_u. Since $[-u, u]$ is complete in E, the space (E_u, p) is a Banach space and hence, an AM-space with unit u. We summarize:

PROPOSITION 1.4: *If E is any Banach lattice and $u \in E_+$ then, under the norm whose closed unit ball is $[-u, u]$, E_u is an AM-space with unit u.*

A compact space K is called *extremally disconnected* (or *Stonian*) if the closure $\overline{U}$ of each open subset $U \subset K$ is open. (The space K is called *quasi-Stonian* if the closure of each open F_σ-subset is open.) Stonian spaces (except in the trivial case of a finite K) rather defy the imagination; yet they are very useful in the theory of Boolean algebras (cf. Part IV), hence in measure theory and also in the theory of Banach lattices. We shall have need of the following relationship.

PROPOSITION 1.5: *A Banach lattice $C(K)$, K compact, is order complete if and only if K is Stonian.*

Proof: Suppose $C(K)$ is order complete and $U \subset K$ is open. If $U \neq \emptyset$, the family A of all functions $f \in [0, e]$ and living on U, is directed ($\leq$) and bounded above by e: by hypothesis, $f_0 := \sup A$ exists in $C(K)$. Now f_0 must be a function such that $f_0(s) = 1$ if $s \in U$ and $f_0(s) = 0$ if $s \notin \overline{U}$. Since f_0 is continuous, f_0 must be the characteristic function of $\overline{U}$ and so $\overline{U}$ must be open.

Conversely, let K be Stonian and denote by A a directed ($\leq$), order bounded subset of $C(K)$. If g is the numerical supremum of A (that is, if $g(t) := \sup\{f(t): f \in A\}$ for all $t \in K$), we define the *upper limit function* g_u of g by

$$g_u(s) := \inf_{U(s)} \sup_{t \in U(s)} g(t) \quad (s \in K),$$

where the infimum is taken over all neighborhoods $U(s)$ of $s \in K$. It is easy to see that g_u is upper semi-continuous and that $f \leq g_u \leq k$ for any $f \in A$ and any upper bound k of A in $C(K)$. The Stonian property of K is now used to show that g_u is lower semi-continuous as well (for details, see [16, II.7.7]). Thus $g_u \in C(K)$ and hence, $g_u = \sup A$; that is, $C(K)$ is order complete. □

Let us also record the following property of Stonian spaces.

PROPOSITION 1.6: *If K is a Stonian space and $G \neq \emptyset$ is an open subset of K, then each bounded continuous function $f: G \to \mathbf{R}$ has a continuous extension $\bar{f}: K \to \mathbf{R}$.*

Since $\overline{G}$ is open-and-closed, we may as well assume that G is dense in K. The proof is along lines similar to those of the second part of the preceding proof; this time we consider the *lower limit function* of f,

$$f_l(s) := \sup_{U(s)} \inf_{t \in G \cap U(s)} f(t) \quad (s \in K),$$

the supremum being taken over all neighborhoods $U(s)$ of $s \in K$. Now f_l maps K into $\mathbf{R}$, agrees with f on G, and is lower semi-continuous. By definition of f_l and with obvious notation we obtain, for any real α and natural numbers $m > n$, the relation $[f \leq \alpha - n^{-1}]^- \subset [f_l \leq \alpha - n^{-1}] \subset [f < \alpha - m^{-1}]^-$. Hence we have

$$[f_l < \alpha] = \bigcup_{n=1}^{\infty} [f_l \leq \alpha - n^{-1}] = \bigcup_{n=1}^{\infty} [f \leq \alpha - n^{-1}]^-$$
$$= \bigcup_{n=1}^{\infty} [f < \alpha - n^{-1}]^-.$$

Now by the continuity of f on G and the Stonian property of K, the set $[f<\alpha-n^{-1}]^-$ is open for each natural n and, hence, f_l is upper semi-continuous. This proves that f_l is the desired continuous extension of f.

2. Spaces of Integrable Functions. For the theory of Banach lattices, next in importance to spaces of continuous functions (§1) are spaces of integrable functions; in particular, the spaces $L^1(\mu)$. By a measure space (X,Σ,μ) we understand a non-void set X, a σ-ring Σ of subsets of X, and a countably additive function $\mu:\Sigma\to\overline{\mathbf{R}}_+$ ($=[0,+\infty]$). As usual, $L^p(\mu)$ (or $L^p(\mu,X)$) denotes the Banach space of equivalence classes (modulo μ-null functions) of p-integrable functions under the norm $\|f\|_p:=(\int|f|^p d\mu)^{1/p}$ ($1\leqslant p<\infty$), and $L^\infty(\mu)$ denotes the Banach space of equivalence classes of essentially bounded Σ-measurable functions under the norm $\|f\|_\infty:=\operatorname{ess\,sup}_{t\in X}|f(t)|$. (For an easily readable introduction to measure theory and the spaces L^p, see Bartle [2].) As remarked above (Part I, §1), the spaces $L^p(\mu)$ are Banach lattices under their natural orderings.

From the definition of the norm in L^p it is at once clear that $f\perp g$ implies $\|f+g\|^p=\|f\|^p+\|g\|^p$ if $1\leqslant p<\infty$; it is our purpose in this section to show that this characterizes L^p-spaces among Banach lattices. We define a Banach lattice E to have *p-additive norm* ($1\leqslant p<\infty$) if $x\wedge y=0$ in E implies $\|x+y\|^p=\|x\|^p+\|y\|^p$.

PROPOSITION 2.1: *If E has p-additive norm ($1\leqslant p<\infty$) then the norm of E is order continuous; in particular, E is order complete.*

The proof results immediately from Theorem I.4.4; in fact, if (x_n) is an orthogonal sequence contained in $[-x,x]\in E$, say, then $\sum_{\nu=1}^n\|x_\nu\|^p\leqslant\|x\|^p$ for all $n\in\mathbf{N}$; hence (x_n) is a null sequence. (It should be noted that for $p=1$, the assertion is easily verified directly, without recourse to Theorem I.4.4.)

An orthogonal family $\{u_\alpha:\alpha\in A\}$ in a Riesz space E is called maximal if $|x|\wedge|u_\alpha|=0$ for all $\alpha\in A$, implies that $x=0$; by Zorn's lemma, every Riesz space contains maximal orthogonal families. If a Riesz space E contains a maximal orthogonal family which is a singleton $\{u\}$, then u is called a *weak order unit* (or a *Freudenthal unit*) of E. We note that if E is Archimedean and contains a weak order unit u, then $x = \sup_n(x \wedge nu)$ for each $x \in E_+$ (see the proof of Proposition I.4.5).

THEOREM 2.2: *If a Banach lattice has p-additive norm, then E is Riesz and norm isomorphic to $L^p(\mu)$ for a suitable measure space (X, Σ, μ).*

Proof: (a) First we consider the case that E has a weak order unit u. By Theorem 1.3 and Proposition 1.4, the principal ideal E_u is an AM-space $C(K)$ (K compact). Since E is order complete by Proposition 2.1, E_u is order complete and so K is Stonian (1.5). Let **Q** denote the Boolean algebra of open-and-closed subsets of K, and let L denote the linear span of the set of characteristic functions $\{\chi_U : U \in \mathbf{Q}\}$. Since L is a (subalgebra and) Riesz subspace of $C(K)$ satisfying the hypotheses of Lemma 1.2, L is dense in $C(K)$ for the uniform norm. We define a set function μ_0 on **Q** by $\mu_0(U) := \|\chi_U\|^p$ (norm of χ_U in E); by the p-additivity of the norm of E, μ_0 is additive and (trivially) countably additive on **Q**. By the Carathéodory extension theorem (see [2, Chapter 9]), μ_0 can be extended to a measure μ on the σ-algebra Σ generated by **Q** (which is the σ-algebra of Baire subsets of K). Now if $f \in L$ then $f = \sum_{i=1}^n \alpha_i \chi_{U_i}$ for a suitable disjoint family $\{U_i : i = 1, \ldots, n\}$ in **Q**; since E has p-additive norm, then

$$\|f\|^p = \sum_{i=1}^n |\alpha_i|^p \|\chi_{U_i}\|^p = \int |f|^p d\mu.$$

This shows that on L, the norm induced by E agrees with the norm of $L^p(\mu, K)$. Now L is uniformly dense in $C(K)$, and $C(K) \cong E_u$ is dense in E by Proposition I.4.5. On the other hand, L is dense in $L^p(\mu, K)$ and so E is isomorphic to $L^p(\mu, K)$.

(b) In the general case let $\{u_\alpha : \alpha \in A\}$ denote a maximal orthogonal subset of E. (Note that if E has no weak order units, each such set is uncountable.) We can assume that $u_\alpha \in E_+$ for all α. By Proposition 2.1 and Proposition I.4.5, the closed ideal E_α generated by $\{u_\alpha\}$ is a projection band, and u_α is a weak order unit of E_α ($\alpha \in A$). By (a) above, E_α is isomorphic to $L^p(\mu_\alpha, K_\alpha)$, where K_α denotes a compact Stonian space ($\alpha \in A$). Let X denote the locally compact space which is the disjoint union $\bigcup_\alpha K_\alpha$, endowed with the topology for which a set U is open iff $U \cap K_\alpha$ is open in K_α for each $\alpha \in A$. Now let $x \in E_+$ and let x_α denote the component of x in the projection band E_α. Then we have $x = \sup_\alpha x_\alpha$; indeed, $y := \sup_\alpha x_\alpha$ exists by the order completeness of E and $x - y \perp x_\alpha$ for all $\alpha \in A$, so $y = x$. (We note that the order continuity of the norm of E implies that $x = \lim_H \Sigma_{\alpha \in H} x_\alpha$ along the net of all finite subsets $H \subset A$; moreover, for given $x \in E$, only a countable number of components x_α can be $\neq 0$, by the p-additivity of the norm of E.) So each $x \in E$ can be viewed as the function on X whose restriction to K_α equals $x_\alpha \in E_\alpha \cong L^p(\mu_\alpha, K_\alpha)$ ($\alpha \in A$), and we have, by orthogonality of the x_α, the relation

$$\|x\|^p = \sum_\alpha \int |x_\alpha|^p d\mu_\alpha.$$

Let Σ denote the σ-ring of subsets of X generated by $\bigcup_\alpha \Sigma_\alpha$, where Σ_α denotes the Baire field of K_α; then each $S \in \Sigma$ is a disjoint countable union $\bigcup_i S_{\alpha_i}$ where $S_{\alpha_i} \in \Sigma_{\alpha_i}$ ($i \in \mathbf{N}$). Defining $\mu(S) := \sum_{i=1}^\infty \mu_{\alpha_i}(S_{\alpha_i})$ yields a measure μ and measure space (X, Σ, μ) for which E is clearly isomorphic to $L^p(\mu)$. □

COROLLARY 1: *If E is a Banach lattice with p-additive norm, then $\|x + y\|^p \geq \|x\|^p + \|y\|^p$ for all $x, y \in E_+$.*

In the special case $p = 1$, p-additivity is equivalent to the condition

(L) $\qquad \|x + y\| = \|x\| + \|y\| \qquad (x, y \in E_+),$

because $\|x + y\| \leq \|x\| + \|y\|$ is always true (triangle inequality). The equality (L) signifies that the norm of E is additive on E_+; because of the following corollary of Theorem 2.2, Banach lattices satisfying (L) are called *abstract Lebesgue spaces* (AL-*spaces*).

COROLLARY 2 (Kakutani): *Every AL-space is Riesz and norm isomorphic to $L^1(\mu)$ for a suitable measure space (X, Σ, μ).*

The place of the spaces $L^\infty(\mu)$ in this theory will be discussed in the following section, where we will encounter a representation of AL-spaces related to the Radon-Nikodým theorem.

3. Duality of AM- and AL-Spaces. Dixmier's Theorem. The AM- and AL-spaces introduced in the preceding sections are dual to each other in the following sense.

PROPOSITION 3.1: *The dual of an AM-space is an AL-space, and the dual of an AL-space is an AM-space with unit.*

Proof: Let E be any AM-space and let $f, g \in E'_+$. For each real $\epsilon > 0$ there exist elements $x, y \in E_+$, $\|x\| = \|y\| \le 1$, such that $f(x) > \|f\| - \epsilon$ and $g(y) > \|g\| - \epsilon$ (cf. Part I, §3, (2)). By condition (M) (§1), the element $z := x \vee y$ satisfies $\|z\| \le 1$ and we obtain $\|f+g\| \ge f(z) + g(z) \ge f(x) + g(y) > \|f\| + \|g\| - 2\epsilon$; hence, $\|f+g\| \ge \|f\| + \|g\|$. Since $\|f+g\| \le \|f\| + \|g\|$ by the triangle inequality, condition (L) (§2) is fulfilled and E' is an AL-space.

If E is an AL-space, then the norm is additive and positive homogeneous on E_+; hence, its linear extension to E is a positive linear form e given by $e(x) = \|x^+\| - \|x^-\|$; also $\|x\| = e(|x|)$ for all $x \in E$. Thus if $f \in E'$ and $\|f\| \le 1$, then we have $|f(x)| \le \|x\| = e(|x|)$ so that $|f| \le e$. Therefore, $[-e, e]$ is the closed unit ball of E', and E' is an AM-space with unit e. □

From I.3.3 Corollary, 1.3, and 3.1, we obtain the following rather weak (but occasionally useful) corollary.

COROLLARY: *Every AM-space is isomorphic to a normed Riesz subspace of some $C(K)$ (K compact).*

The reader is cautioned not to believe that every AM-space with unit and every AL-space is a dual Banach lattice; for example, $C[0,1]$ is not because it is not order complete, and $L^1[0,1]$ (with Lebesgue measure) is not because its positive cone contains no extreme rays. The prime purpose of this section is to characterize those compact spaces K for which $C(K)$ is a dual Banach lattice (or, as it turns out, a dual Banach space). We point out that Theorem 3.3, Corollary 1, below was instrumental in the proof of Theorem I.4.4. We begin with a purely measure theoretic result.

PROPOSITION 3.2: *Let K denote a compact Stonian space and suppose that μ is a bounded, regular Borel measure on K vanishing on all closed rare subsets of K and such that $\mu(U) > 0$ for each open set $U \neq \varnothing$. Then $C(K)$ is Riesz and norm isomorphic with $L^\infty(\mu, K)$.*[†]

Proof: We show that for each Borel set $A \subset K$, there exists an open-and-closed set $V \subset K$ such that $\mu(V \Delta A) = 0$. By the regularity of μ, there exist open sets $U_n \supset A$ satisfying $\mu(U_n \setminus A) \leq n^{-1}$ ($n \in \mathbb{N}$); since $\overline{U_n} \setminus U_n$ is rare, we conclude that $\mu(\overline{U_n} \setminus A) \leq n^{-1}$. The set $V_0 := \bigcap_{n=1}^{\infty} \overline{U_n}$ is closed, contains A, and satisfies $\mu(V_0 \setminus A) = 0$. The set $W := K \setminus V_0$ is open, and again $\mu(\overline{W}) = \mu(W)$ by the hypothesis on μ. Since K is Stonian, $\overline{W}$ is open-and-closed, so $V := K \setminus \overline{W}$ is open-and-closed and satisfies $\mu(V \Delta A) = 0$. The characteristic function χ_V is continuous and, since μ is strictly positive on non-void open sets, the unique continuous function in its equivalence class in $L^\infty(\mu, K)$. Therefore, if L denotes the dense (cf. 1.2) Riesz subspace of all functions in $C(K)$ with finite range, then L is also a dense Riesz subspace of $L^\infty(\mu, K)$. It is clear that the identity map of L extends to a Riesz and norm isomorphism of $C(K)$ onto $L^\infty(\mu, K)$. □

THEOREM 3.3 (Dixmier): *Let K be a compact Stonian space and suppose that the AL-space H of all order continuous linear forms on $C(K)$ separates $C(K)$. Then $C(K)$ is Riesz and norm isomorphic to H'; in other words, H is a predual of $C(K)$. Moreover, there exists a Borel measure ν on K such that $H \cong L^1(\nu, K)$ and $C(K) \cong L^\infty(\nu, K)$.*

Proof: By Proposition 3.3 of Part I, H is a Riesz subspace of $C(K)'$ and it is clear that H is norm closed; hence by Proposition 3.1, H is an AL-space. Now let $\{\nu_\alpha : \alpha \in A\}$ denote a maximal orthogonal subset of H_+. In the usual way, we consider each ν_α to

---

[†]The subsequent proof shows an analogous result to hold for a Baire measure on K.

be a regular Borel measure on K (Riesz representation theorem). Let K_α denote the carrier of ν_α (that is, let K_α denote the complement of the largest open set $U_\alpha \subset K$ such that $\nu_\alpha(f)=0$ whenever $[f\neq 0]^- \subset U_\alpha$, $f\in C(K)$). Since K is Stonian, $C(K)$ is order complete by Proposition 1.5 and $\bar{f}_\alpha := \sup C_\alpha$ exists in $C(K)$, where C_α denotes the set of all $f\in[0,e]$ satisfying $f(W)\subset\{0\}$ for some neighborhood W of K_α. The function $\bar{f}_\alpha$ must be the characteristic function of $\bar{U}_\alpha$; but ν_α is order continuous, so $\nu_\alpha(\bar{f}_\alpha)=0$ and thus $U_\alpha = \bar{U}_\alpha$ is open-and-closed. Moreover, it is easily verified that $K_\alpha \cap K_\beta = \varnothing$ whenever $\alpha, \beta \in A$ and $\alpha \neq \beta$.

Now each pair (K_α, ν_α) satisfies the hypothesis of Proposition 3.2, because by order continuity, ν_α vanishes on rare Borel sets, so that $C(K_\alpha) \cong L^\infty(\nu_\alpha, K_\alpha)$. As in part (b) of the proof of Theorem 2.2, we let X denote the (locally compact) direct sum of the compact spaces K_α ($\alpha \in A$) and let ν denote the measure on X induced by the family (ν_α). (Precisely, we take Σ to be the σ-algebra of all unions $S = \bigcup_\alpha S_\alpha$, S_α any Borel set in K_α, and define $\nu: \Sigma \to \bar{\mathbf{R}}_+$ by $\nu(S) = \Sigma \nu_\alpha(S_\alpha)$.) From the same proof it is then clear that $H \cong L^1(\nu, X)$, each $f \in L^1(\nu, X)$ being a countable sum $f = \Sigma_i f_{\alpha_i}$ where $f_\alpha \in L^1(\nu_\alpha, K_\alpha)$ and $\|f\| = \Sigma_i \|f_{\alpha_i}\|$. Thus H is an $l^1(A)$-direct sum of the AL-spaces $L^1(\nu_\alpha, K_\alpha)$ ($\alpha \in A$). Since each $L^1(\nu_\alpha, K_\alpha)$ has the dual $L^\infty(\nu_\alpha, K_\alpha) \cong C(K_\alpha)$ (3.2), each continuous linear form g on H is given by a uniformly bounded family $(g_\alpha)_{\alpha \in A}$, where $g_\alpha \in C(K_\alpha)$ and $\|g\| = \sup_\alpha \|g_\alpha\|$; conversely, each such family defines a unique $g \in H'$. Now g defines, in an obvious manner, a bounded continuous function on $X = \bigcup_\alpha K_\alpha$. Since H separates $C(K)$ by hypothesis, X must be dense in K; therefore, by Proposition 1.6, the function on X defined by g has a unique continuous extension $\bar{g}: K \to \mathbf{R}$. It follows that the dual of H is Riesz and norm isomorphic to $C(K)$ as asserted (K being, in fact, the Stone-Čech compactification of X).

Finally, the measure ν on (X, Σ) defined above can be extended to a Borel measure on K by letting $\nu(K\setminus X) = 0$; clearly, then, we have $H \cong L^1(\nu, K)$ and $C(K) \cong L^\infty(\nu, K)$. □

COROLLARY 1: *Let $C(K)$ be order complete and let N be a separating family of order continuous linear forms on $C(K)$. Then the closed unit ball of $C(K)$ is $\sigma(C(K), N)$-compact.*

Proof: By the preceding theorem, the closed unit ball $[-e,e]$ of $C(K)$ is $\sigma(C(K),H)$-compact. Since $\sigma(C(K),N)$ is a coarser Hausdorff topology (N being assumed separating), then $[-e,e]$ is compact for this topology as well. □

It is customary to call a compact Stonian space K *hyperstonian* if $C(K)$ is separated by its order continuous linear forms.

COROLLARY 2 (Dixmier-Grothendieck): *Let K be compact. For $C(K)$ to be isometrically isomorphic to a dual Banach space, it is necessary and sufficient that K be hyperstonian.*

Proof: The sufficiency of the condition is clear from Theorem 3.3. Conversely, if $C(K) \cong G'$ for some Banach space G, then $[-e,e]$ is $\sigma(C(K),G)$-compact. Since $C(K)_+ \cap [-e,e] = [0,e] = \frac{1}{2}([-e,e]+e)$, the Krein-Šmulian theorem implies that the positive cone $C(K)_+$ is $\sigma(C(K),G)$-closed. Hence by weak* compactness, $[-e,e]$ is order complete; since e is an order unit of $C(K)$, $C(K)$ is order complete and hence, K is Stonian (1.5). Now each element of G defines a weak* continuous, hence order continuous, linear form on $C(K)$, and so K is hyperstonian. □

It is now easy to obtain a second representation theorem for AL-spaces.

THEOREM 3.4: *Suppose L is an AL-space. Then there exists a compact (necessarily hyperstonian) space K, unique to within homeomorphism, such that L is Riesz and norm isomorphic to the AL-space of all bounded, regular Borel measures on K vanishing on rare Borel sets.*

Proof: By Proposition 3.1 and Theorem 1.3, the dual of L is a space $C(K)$ where K is compact hyperstonian. Let $\tilde{L} := C(K)'_{oc}$; it is not difficult to verify that if μ is a bounded, regular Borel measure on K (or equivalently, by the Riesz representation theorem, a continuous linear form on $C(K)$), then μ is order continuous on $C(K)$ if and only if $|\mu|(A) = 0$ for each rare Borel set $A \subset K$ (consider the characteristic function of a closed nowhere dense set

B as the infimum of the net of all $f \in C(K)_+$ satisfying $f(B) \subset \{1\}$). Thus we have $L \subset \tilde{L}$ and, by Theorem 3.3, $\tilde{L}' \cong C(K)$; but L separates its dual $C(K)$ and hence must be weakly dense in $\tilde{L}$. Since L is convex and norm closed in $\tilde{L}$, it follows that $L = \tilde{L}$.

Finally, if K_1 is any compact space with the desired property, then $L \cong C(K)'_{oc}$ by the preceding, and the adjoint of the identity map $L \to L$ defines an isometric isomorphism of Banach lattices $C(K) \to C(K_1)$. From Theorem 1.3 it now follows that this isomorphism induces a homeomorphism of K onto K_1. □

4. Banach Lattices with Topological Units. Many approaches have been made to represent abstract (Archimedean) Riesz spaces as spaces of real- or extended real-valued functions; an extensive account of this subject is given by Luxemburg-Zaanen [12]. The representation theory for Banach lattices, whose first results will be given in this section, is of rather recent origin (Lotz, Davies, 1968; for a detailed account, see [16, Chapter 3]) and very intimately related to Kakutani's representation theorem (1.3). In the following, order units are replaced by their topological analogues: An element $u \in E_+$ of a Banach lattice E is called a *topological unit* of E (or a *quasi-interior point* of E_+) if the principal ideal E_u is dense in E. The representation we seek rests on the fact that in the presence of topological units, E contains dense ideals of the form $C(K)$ (K compact) (cf. 1.3 and 1.4).

Topological units exist in every separable Banach lattice [16, II.6.2]. From our next result it follows that in a Banach lattice with order continuous norm, each weak order unit is a topological unit.

PROPOSITION 4.1: *Let E be any Banach lattice. An element $u \in E_+$ is a topological unit if and only if $x = \lim_n (x \wedge nu)$ for each $x \in E_+$. Moreover, if u and v are topological units of E, then the AM-spaces E_u and E_v (cf. 1.4) are canonically isomorphic.*

Proof: Clearly, if $x = \lim_n (x \wedge nu)$ for all $x \in E_+$, then E_u is dense in E since $x_n := x \wedge nu \in E_u$. Conversely, if $\bar{E}_u = E$, then for each $x \in E_+$ there exists a sequence (y_k) in $(E_u)_+$ with limit x. Letting $z_k := (y_1 \vee \cdots \vee y_k) \wedge x$, we have (z_k) increasing and, by continuity of the lattice operations, $x = \lim_k z_k$. Since $z_k \in E_u$ for all

k, there exists an integer-valued function $k \mapsto n(k)$ such that $z_k \leq n(k)u$ for all k. Therefore, $z_k \leq x_{n(k)} \leq x$ and so $\lim_k x_{n(k)} = x$; since (x_n) is increasing, it follows that $\lim_n x_n = x$.

To prove the second assertion, we note that if E_u and E_v are dense ideals, then so is $E_u \cap E_v = E_{u \wedge v}$; hence $u \wedge v$ is a topological unit of E whenever u and v are. Thus, to prove the existence of a canonical Riesz isomorphism of E_u onto E_v, we may and will assume that $v \leq u$; by Theorem 1.3 we can identify E_u with $C(K_u)$, K_u compact. Considering E_v as an ideal of $C(K_u)$, we next observe that the set $V := \{s \in K_u: v(s) > 0\}$ is dense in K_u for, v being a topological unit of E, v is a weak order unit of $C(K_u)$. Thus the mapping $\phi_{u,v}: E_u \to E_v$ defined by $h \mapsto vh$ (pointwise multiplication of functions in $C(K_u)$) is injective; if we can show it to be surjective, it will be the desired Riesz isomorphism. To this end, let $g \in E_v$ be given and assume for convenience that $0 \leq g \leq v$; then $k := v^{-1}g$ defines a continuous function on V, and we have to show that k extends to a continuous real function on all of K_u. Consider the strictly positive functions $v_n := n^{-1}(u \vee nv)$; the sequence (v_n) is decreasing with pointwise infimum v on K_u. Let $h_n := v_n^{-1}g$ ($n \in \mathbb{N}$), then $|h_n - h_m| \leq |v_n^{-1}v - v_m^{-1}v|$ for all $m, n \in \mathbb{N}$. Since v is a topological unit of E, then $v_n^{-1}v = u \wedge nv$ defines a Cauchy sequence in E. It follows that (h_n) is Cauchy in E and converges to an element $h \in E_u$ satisfying $g = vh$. □

Thus if u,v are topological units of E and $E_u \cong C(K_u)$, $E_v \cong C(K_v)$ are the corresponding Kakutani representations, then, by the corollary of Theorem 1.3, K_u and K_v are (canonically) homeomorphic; any representative of this class of pairwise homeomorphic compact spaces is called the *structure space* K_E of E. Except when E is an AM-space with unit, it is impossible to consider K_E as the space of maximal ideals, or equivalently, as the space of (appropriately normalized) real Riesz homomorphisms of E. This deficiency will now be remedied by extending the notion of real Riesz homomorphism, as follows.

We consider the one-point compactification $\overline{\mathbb{R}}_+ = \{\mathbb{R}_+, \infty\}$ as a lattice in the obvious way ($\infty = \sup \mathbb{R}_+$), and as an abstract cone by defining $a + \infty = \infty + a = \infty$ for all $a \in \overline{\mathbb{R}}_+$ and by defining $0\infty = 0$, $\lambda\infty = \infty$ for all $0 < \lambda \in \mathbb{R}_+$. If E is any Riesz space, then a *valuation of E* is a mapping $\phi: E \to \overline{\mathbf{R}}_+$ satisfying these axioms:

(V$_1$) $\phi(x+y) = \phi(x) + \phi(y)$ for all $x, y \in E_+$,

(V$_2$) $\phi(x \wedge y) = \phi(x) \wedge \phi(y)$ for all $x, y \in E_+$,

(V$_3$) $\phi(\lambda x) = |\lambda| \phi(|x|)$ for all $x \in E, \lambda \in \mathbf{R}$.

From these axioms we obtain $\phi(0) = 0$ and, for all $x, y \geq 0$, $\phi(x \vee y) = \phi(x) \vee \phi(y)$. The valuation ϕ is called *u-normalized* if $\phi(u) = 1$ for some $u \in E_+$. Every Riesz homomorphism $f: E \to \mathbf{R}$ gives rise to a valuation $x \mapsto \phi(x) := f(|x|)$; these are finite (and the only finite) valuations of E. The following extension theorem (Lotz, 1968) shows how to construct other valuations.

PROPOSITION 4.2: *Let E be any Riesz space, and let $0 < u \in E$. Every u-normalized valuation[†] of E_u can be extended to a unique valuation of E.*

Proof: Let ϕ denote a valuation of E_u satisfying $\phi(u) = 1$. We define an extension $\overline{\phi}$ of ϕ to E by $\overline{\phi}(x) := \sup\{\phi(y): y \in [0, |x|] \cap E_u\} \in \overline{\mathbf{R}}_+$ for all $x \in E$. To prove (V$_1$), let $x, y \geq 0$ and observe that $[0, x+y] \cap E_u = [0, x] \cap E_u + [0, y] \cap E_u$ by (D) of Proposition I.2.1. To verify (V$_2$) it now suffices to show that $\overline{\phi}(x \wedge y) = 0$ whenever $x \wedge y = 0$. Defining $I := E_x \in E_u$ and $J := E_y \cap E_u$, we have $I \cap J = \{0\}$ and hence, since ϕ is finite on E_u and so $\phi^{-1}(0)$ is a maximal ideal of E_u, at least one of the relations $\phi(I) = \{0\}$ or $\phi(J) = \{0\}$ holds; thus $\overline{\phi}(x \wedge y) = 0$. Finally, $\overline{\phi}$ is unique: Suppose χ, ψ are (u-normalized) valuations of E extending ϕ such that $\chi(x) \neq \psi(x)$ for some $x \in E_+$; by (V$_3$) we can assume that $\chi(x) < 1 < \psi(x)$. Then by (V$_2$), $\chi(x \wedge u) = \chi(x) < \psi(u) = \psi(x \wedge u)$, which is contradictory, since $x \wedge u \in E_u$. □

Suppose now that u is a topological unit of a Banach lattice E; considering the set V_u of u-normalized valuations as a (closed) subspace of the compact product $\overline{\mathbf{R}}_+^{E_+}$, V_u is a compact space under the weak topology $\sigma(V_u, E_+)$. On the other hand, E_+ thus

---

[†] For example, if E is a Banach lattice and $E_u \cong C(K_u)$, the u-normalized valuations of E_u are precisely $f \mapsto |f(t)|$, $t \in K_u$.

appears as a cone of continuous numerical functions $V_u \to \overline{\mathbf{R}}_+$. (We recall that a *numerical function* is a function into the two-point compactification $\overline{\mathbf{R}}$ of $\mathbf{R}$.) Moreover, by the preceding extension theorem, restricting each $\phi \in V_u$ to $E_u \cong C(K_u)$ defines a homeomorphism of V_u onto K_u; here K_u is identified, in the spirit of Kakutani's theorem (1.3), with the $\sigma(E'_u, (E_u)_+)$-compact set of all u-normalized valuations of E_u. It is on this identification of V_u with K_u that the following representation theorem is based.

THEOREM 4.3: *Every Banach lattice E with topological units is Riesz isomorphic to a Riesz space $\hat{E}$ of continuous numerical functions on the structure space K_E, each of which is infinite only on some rare subset of K_E. The space $\hat{E}$ contains $C(K_E)$ as a dense ideal, and this property determines K_E to within homeomorphism.*

Note: It is understood in this theorem that the lattice operations of $\hat{E}$ agree with the pointwise operations and that the algebraic operations agree with the pointwise operations on sets of finiteness. The density assertion refers to the topology of $\hat{E}$ transferred from E.

Proof of (4.3): Let u denote any topological unit of E, consider $E_u \cong C(K_u)$, and let K_E be represented by K_u. By the preceding observations, each $x \in E_+$ can be viewed as a continuous function $\hat{x}: \phi \mapsto \phi(x)$ of $V_u \cong K_u \to \overline{\mathbf{R}}_+$. By defining $\hat{x}(\phi) = \phi(x^+) - \phi(x^-)$ for all $x \in E$ and $\phi \in V_u$, $\hat{x}$ becomes a continuous numerical function on V_u (note that by (V_2), at most one of $\phi(x^+)$, $\phi(x^-)$ is infinite). Moreover, it is clear that $x \mapsto \hat{x}$ maps E_u isomorphically onto $C(K_u)$. Next we observe that this mapping is injective: If $\phi(x_1) = \phi(x_2)$ for all $\phi \in V_u$ (we can assume that $x_1, x_2 \geq 0$), then $\phi(x_1 \wedge nu) = \phi(x_2 \wedge nu)$ for all ϕ and hence $x_1 \wedge nu = x_2 \wedge nu$ for all $n \in \mathbf{N}$ by the preceding; Proposition 4.1 shows that $x_1 = x_2$. To see that for each $x \in E$ the set $H_x := \{\phi \in V_u : \hat{x}(\phi) = \pm \infty\}$ is rare, assume to the contrary that for some $x > 0$, H_x contains an interior point. By Urysohn's theorem, there exists a function $\hat{y}$, $0 < \hat{y} \in C(V_u)$, supported by H_x. Now $\hat{y}$ is the image of some $y \in E_u$ and (V_2) implies that $\phi(x \wedge ny) = \phi(ny)$ for all $\phi \in V_u$; consequently, $x \geq ny$ for all $n \in \mathbf{N}$ which is contradictory, since the order of E is Archimedean.

It is clear from the definition above that $x \mapsto \hat{x}$ preserves the lattice operations. To show that $(\alpha x + \beta y)\hat{} = \alpha\hat{x} + \beta\hat{y}$ ($\alpha, \beta \in \mathbf{R}$), note that by the preceding, the set U where both $\hat{x}$ and $\hat{y}$ are finite, is a dense open subset of V_u; by (V_1), $\alpha\hat{x}(\phi) + \beta\hat{y}(\phi) = (\alpha x + \beta y)\hat{}(\phi)$ for $\phi \in U$. Hence on U, $\alpha\hat{x} + \beta\hat{y}$ must be the restriction of $(\alpha x + \beta y)\hat{}$, and this shows $x \mapsto \hat{x}$ to be linear.

Finally, if K is compact and $j: C(K) \to E$ is a Riesz isomorphism onto a dense ideal of E, then $v := j(e)$ (e the unit of $C(K)$) is a topological unit of E and $C(K)$ is Riesz isomorphic to E_v. By Proposition 4.1 and the corollary of Theorem 1.3, K is homeomorphic to K_v and hence to K_E. □

We have included this rather compressed approach to Theorem 4.3, not only because it is the first step toward a simultaneous extension of both of Kakutani's theorems (1.3) and (2.2) (namely, to Banach lattices possessing a topological unit), but also because it will be extremely useful in our later applications (Part IV, §3). Space does not permit any further elaboration. Suffice it to say that both a characterization of the normed structure of E (in terms of measures on the structure space) and an extension to a larger class of Banach lattices (in particular, those having order continuous norm) are perfectly feasible. For details and further results see [16, Chapter III, §§4-6].

PART III. TENSOR PRODUCTS AND EXTENSION OF OPERATORS

1. Norm-Preserving Extension of Linear Operators. The classical Hahn-Banach theorem for normed spaces can be stated as follows: If $j: G_0 \to G$ is an isometric injection of normed spaces and if $f_0: G_0 \to \mathbf{R}$ is a bounded linear form, then there exists a linear form $f: G \to \mathbf{R}$ of norm $\|f\| = \|f_0\|$ which makes the following diagram commute:

This formulation suggests the question: What Banach spaces H can replace $\mathbf{R}$ in the diagram, G_0 and G being left arbitrary? It is easy to see, and must have been observed early, that each space $l^\infty(A)$ (A any index set) will do; it suffices indeed to extend each coordinate form separately. However, it was not known until the early fifties (Goodner, Nachbin, 1951; Kelley, 1952) that a Banach space H has the extension property in question if and only if H is isometrically isomorphic to some space $C(K)$, where K is compact Stonian. Sufficiency is most simply proved by observing that the classical proof of the Hahn-Banach theorem remains formally and actually in force if $\mathbf{R}$ is replaced by the order complete Riesz space $C(K)$, and the sublinear functional p is replaced by the sublinear function $G \to C(K)$ defined by $p(x) := \|x\|e$ (e the unit of $C(K)$) (cf. [16, II.7.10]).

We shall, however, consider a different approach to this problem, giving us more insight and leading us to new extension theorems for positive and regular maps (Part I, §5) between Banach lattices. Our starting point is the following equivalent of the Hahn-Banach theorem.

PROPOSITION 1.1: *Let G, H be normed spaces with Banach duals G', H', and let $j: G \to H$ be a continuous linear map. Then j is an isometry if and only if $j': H' \to G'$ is a strict metric homomorphism.*

Proof: We recall that a linear map $T: G \to H$ is called a *metric homomorphism* if T maps the open unit ball of G onto the open unit ball of H; T is called *strict* if it maps the closed unit ball of G onto the closed unit ball of H.

Now if j is an isometry, then $j(G)$ can be identified with a normed subspace of H and the Hahn-Banach theorem shows that each $x' \in G'$, $\|x'\| \leq 1$, equals $j'(y')$ (j' the adjoint of j) for some $y' \in H'$, $\|y'\| \leq 1$; thus j' is a strict metric homomorphism. Conversely, if j' is a strict metric homomorphism, then $x' = j'(y')$ for given x' and suitable y' in the respective dual unit balls. Therefore,

$$\|j(x)\| = \sup_{\|y'\| \leq 1} \langle j(x), y' \rangle = \sup_{\|y'\| \leq 1} \langle x, j'(y') \rangle = \sup_{\|x'\| \leq 1} \langle x, x' \rangle = \|x\|,$$

and hence j is an isometry. □

Let us now look at the continuous bilinear forms $\phi\colon G\times H\to \mathbf{R}$; it is immediate that the vector space of all these forms is a Banach space $\mathfrak{B}(G,H)$ under the norm $\|\phi\|:=\sup\{|\phi(x,y)|\colon \|x\|\leq 1, \|y\|\leq 1\}$. On the other hand, each $\phi\in\mathfrak{B}(G,H)$ gives rise to a linear form $\hat{\phi}\colon G\otimes H\to\mathbf{R}$; if $\chi\colon G\times H\to G\otimes H$ denotes the natural bilinear map and W the convex hull of $\{x\otimes y\colon \|x\|\leq 1, \|y\|\leq 1\}$, then $\hat{\phi}$ is bounded by $\|\phi\|$ on W. The set W is a convex, circled, absorbing subset of $G\otimes H$ whose gauge function is a norm, called the π-norm of $G\otimes H$; it is clear from the preceding that each linear form $\hat{\psi}$ on $G\otimes H$, bounded on W, defines a bilinear form $\psi\in\mathfrak{B}(G,H)$ by virtue of $\psi(x,y):=\hat{\psi}(x\otimes y)$. Thus under the map $\phi\mapsto\hat{\phi}$, the space $\mathfrak{B}(G,H)$ is isometrically isomorphic with the dual of $G\otimes_\pi H$ (or of its completion $G\tilde{\otimes}_\pi H$).

The other relevant aspect of $\mathfrak{B}(G,H)$ is operator-theoretic. If $\phi\in\mathfrak{B}(G,H)$, let us denote by $T_\phi x$ the continuous linear form $y\mapsto\phi(x,y)$ on H; by the (joint) continuity of ϕ, the dependence of $T_\phi x$ on x is (linear and) continuous. Thus $T_\phi\in\mathcal{L}(G,H')$ and it is quickly seen that the identity

$$\phi(x,y)=\langle T_\phi x, y\rangle \qquad (x\in G, y\in H),$$

establishes an isometric isomorphism of $\mathfrak{B}(G,H)$ onto the Banach space of bounded linear operators $\mathcal{L}(G,H')$. In particular, $\mathcal{L}(G,H')$ can be identified with the dual of $G\otimes_\pi H$.

The extension property considered above is thus equivalent to a metric property of the π-tensor product, as follows.

PROPOSITION 1.2: *Let G_0, G, H be normed spaces, G_0 denoting a normed subspace of G. The following are equivalent:*

(a) *Every continuous linear operator $T_0\colon G_0\to H'$ has a norm-preserving linear extension $T\colon G\to H'$.*

(b) *The natural imbedding $G_0\otimes_\pi H\to G\otimes_\pi H$ is an isometry.*

Proof: Under the isometric isomorphisms $(G_0\otimes_\pi H)'\cong \mathcal{L}(G_0,H')$ and $(G\otimes_\pi H)'\cong\mathcal{L}(G,H')$ considered above, the adjoint of the natural (or canonical) imbedding $j\colon G_0\otimes_\pi H\to G\otimes_\pi H$ is precisely the map j' which orders to each $T\colon G\to H'$ its restriction $T_0:=T|G_0$. Thus the equivalence of (a) and (b) follows at once from Proposition 1.1. □

The definition of the π-norm on $G \otimes H$ given above has its analytic equivalent in the formula

$$\|u\|_\pi = \inf \sum_i \|x_i\| \|y_i\|,$$

where the infimum is taken over all representations $u = \sum_i x_i \otimes y_i$ of $u \in G \otimes H$. The interpretation of the π-norm is easy if one of the factors, H, say, is a Banach space $L^1(\mu)$. We can simplify matters even further by replacing H with the dense subspace H_0 of all simple μ-integrable functions, that is, functions of the form $f = \sum_{i=1}^n \alpha_i \chi_i$ where χ_i denotes the characteristic function of an integrable set A_i and the A_i can be taken disjoint. The tensor product $G \otimes H_0$ then consists of all simple G-valued functions $u = \sum_{i=1}^n x_i \otimes \chi_i$, and it is found at once that

$$\|u\|_\pi = \sum_{i=1}^n \|x_i\| \|\chi_i\| = \int \|u(t)\| d\mu(t)$$

for these functions u. Now if, as above, G_0 is a normed subspace of G, the natural map $G_0 \otimes_\pi H_0 \to G \otimes_\pi H_0$ is an isometry; clearly, then, $G_0 \otimes_\pi L^1(\mu) \to G \otimes_\pi L^1(\mu)$ is an isometry for any AL-space $L^1(\mu)$. (We note in passing that the completion $G \tilde{\otimes}_\pi L^1(\mu)$ can be identified with the Banach space $L_G^1(\mu)$ of all G-valued functions that are *Bochner integrable* with respect to μ.)

Thus we can apply (1.2), (b)$\Rightarrow$(a), when $H = L^1(\mu)$ and hence $H' \cong C(K)$ with K compact hyperstonian (II.3.3, Corollary 2). The more general case, where the target space $C(K)$ is order complete but not a dual space (that is, where K is Stonian but not hyperstonian), can then be reduced to the preceding by the following lemma due to Gleason.

LEMMA 1.3: *Let X, Y be compact spaces and let X be Stonian. If C is a Riesz subspace of $C(Y)$, Riesz and norm isomorphic to $C(X)$, then there exists a contractive projection $C(Y) \to C$ which is a Riesz homomorphism.*

Proof: Let R denote the equivalence relation in Y defined by $s \sim t$ iff $f(s) = f(t)$ for all $f \in C$. The quotient space Y/R is compact and, since C contains the unit e of $C(Y)$, homeomorphic to X

by Theorem II.1.3, Corollary. By Zorn's lemma and the compactness of Y, there exists a minimal closed subspace Y_0 of Y on which the quotient map $q: Y \to Y/R = X$ is still surjective. We claim that the restriction $q_0: Y_0 \to X$ is injective (and hence, by the compactness of Y_0 and X, a homeomorphism). If so, there exists a Riesz and norm isomorphism $C(Y_0) \to C$ which, when composed with the restriction map $C(Y) \to C(Y_0)$, yields a projection $C(Y) \to C$ of the desired kind.

To show that q_0 is injective we first show that q_0 is an open map. In fact, if U is any non-void open subset of Y_0 then, by the minimality of Y_0, $q_0(U)$ cannot be nowhere dense in X. Thus $\overline{q_0(U)}$ contains interior points for each neighborhood of any point $t_0 \in Y_0$; in particular, since q_0 is continuous, $q_0(t_0)$ is in the closure W of $\text{int} q(U)$. Since X is Stonian, W is open and hence a neighborhood of $q_0(t_0)$; thus q_0 is open. Now if q_0 were not injective, there would exist distinct points $t_0, t_1 \in Y_0$ satisfying $q_0(t_0) = q_0(t_1)$; by continuity of q_0, there would exist disjoint open neighborhoods $U(t_0)$, $U(t_1)$ such that $q_0(U(t_1)) \subset q_0(U(t_0))$. Thus we would have $q_0(Y_0 \setminus U(t_1)) = X$ which contradicts the minimality of Y_0. □

After this (seemingly annoying but, as we will see, rewarding) detour we arrive at the initially cited result.

THEOREM 1.4: *Let K be a compact Stonian space. Every continuous linear map from a normed space G_0 into $C(K)$ has a norm-preserving linear extension to any normed space G containing G_0 as a normed subspace.*

Proof: By Proposition II.3.1 and Theorem II.2.2, Corollary 2, the dual H of $C(K)$ is a space $L^1(\mu)$: thus by (1.2) and the subsequent discussion, every $T_0: G_0 \to C(K) \subset H' = C(K)''$ has a norm-preserving extension $\tilde{T}: G \to C(K)''$. Considering $C(K)$ as a normed Riesz subspace of $C(K)''$ (cf. I.3.3, Corollary), by Lemma 1.3 there exists a contractive projection (which is even a Riesz homomorphism) $P: C(K)'' \to C(K)$. Now $T := P\tilde{T}$ is an extension of T_0 as required. □

Conversely, as we have remarked above, a target space having the extension property of Theorem 1.4 is necessarily isometric to some $C(K)$, K Stonian. Thus if (and only if) H is a Banach space such that the natural map $G_0 \otimes_\pi H \to G \otimes_\pi H$ is an isometry for every normed space G and normed subspace G_0, then by Proposition 1.2 $H' \cong C(K)$ and so by Theorem II.3.3, Corollary 2, H is an AL-space. This characterization of AL-spaces is due to Grothendieck [8].

2. Tensor Products of Archimedean Riesz Spaces. In the language of categories, the extension theorem (1.4) simply states that an object in **BS** is *injective* if it is (isomorphic to) a space $C(K)$, where K is compact Stonian; here **BS** is the category of all Banach spaces, the linear contractions serving as the morphisms of the category. It is now natural to ask for injective objects in the category **BL** of all Banach lattices with morphisms the positive contractions; it turns out that the Banach lattices $C(K)$ (K Stonian) are injective in **BL** as well. But in contrast to **BS**, there are many other injective objects in **BL**, notably all AL-spaces, as first shown by Lotz [10] (cf. [16, II.8.9]). We will prove these results and the very general Theorem 4.3 below by a method imitating the approach for Banach spaces outlined in §1; this necessitates the introduction of a tensor product of Archimedean Riesz spaces first constructed by Fremlin [6].

Let E, F, G denote Riesz spaces and let $\psi: E \times F \to G$ be a bilinear map. The map ψ is called *positive* if $\psi(x, y) \geq 0$ whenever $x \geq 0$ and $y \geq 0$, and ψ is called a *Riesz bimorphism* if $x \mapsto \psi(x, y)$ and $y \mapsto \psi(x, y)$ are Riesz homomorphisms for each $y > 0$ and $x > 0$, respectively. The prototype of a Riesz bimorphism is the bilinear form $(f, g) \mapsto \gamma f(s) g(t)$ on $C(X) \times C(Y)$, where X, Y are compact spaces and $s \in X$, $t \in Y$, $\gamma \in \mathbf{R}_+$ are fixed. We begin with an extension and factoring result applying to this special situation.

PROPOSITION 2.1: *Let X, Y, Z denote compact spaces, and let $E_0 \subset C(X)$, $F_0 \subset C(Y)$ denote dense Riesz subspaces, each containing the constant functions. Every positive bilinear map $\phi: E_0 \times F_0 \to C(Z)$ is continuous and hence has a (unique) continuous positive extension $\tilde{\phi}: C(X) \times C(Y) \to C(Z)$; if ϕ is a Riesz bimorphism then* *so is $\tilde\phi$. Moreover, every Riesz bimorphism $\psi: C(X)\times C(Y)\to C(Z)$ can be factored*

$$C(X)\times C(Y) \overset{\chi}{\to} C(X\times Y) \overset{\tau}{\to} C(Z),$$

where χ is the natural bilinear map and τ is a Riesz homomorphism uniquely determined by ψ.

Proof: If we denote by f,g elements of E_0, F_0, respectively, so that $|f|\leq \|f\|e_X$, $|g|\leq \|g\|e_Y$ (e the constant-one function), then positivity of ϕ yields

$$|\phi(f,g)| \leq \phi(|f|,|g|) \leq \|f\|\,\|g\|\phi(e_X, e_Y),$$

which proves that ϕ is continuous. Since $E_0\times F_0$ is dense in $C(X)\times C(Y)$ by hypothesis, ϕ has a unique continuous extension with the desired properties. (Note that ϕ is uniformly continuous on the product $U_0\times V_0$ of the respective unit balls and the latter is dense in $U\times V$.) Clearly, if ϕ is a Riesz bimorphism, then so is $\tilde\phi$.

Now let $\psi: C(X)\times C(Y)\to C(Z)$ be a Riesz bimorphism and fix $r\in Z$. For fixed g, $0<g\in C(Y)$, the map $f\mapsto \langle \psi(f,g),\delta_r\rangle$ (δ_r the evaluation functional or Dirac measure at $r\in Z$) is a real Riesz homomorphism of $C(X)$ and hence, by Theorem II.1.3, Corollary, of the form $f\mapsto \alpha f(s_r)$ for suitable $s_r\in X$ and constant $\alpha\in \mathbf{R}_+$ depending on g and r. On the other hand, for fixed $f>0$, $g\mapsto \alpha$ is a Riesz homomorphism of $C(Y)$ whence $\alpha = \gamma_r g(t_r)$ for suitable $t_r\in Y$ and $\gamma_r\geq 0$. If follows that $\delta_r\circ\psi$ can be factored through $C(X\times Y)$ by virtue of the map $\tau_r: h\mapsto \gamma_r h(s_r, t_r)$ and hence, since $\{\delta_r: r\in Z\}$ separates $C(Z)$, we have $\tau_r = \delta_r\circ \tau$ for a uniquely determined Riesz homomorphism $\tau: C(X\times Y)\to C(Z)$. □

We now consider Archimedean Riesz spaces E, F, G. If $u\in E_+$, the principal ideal E_u generated by $\{u\}$ is an Archimedean Riesz space with order unit u; therefore, the gauge function p_u of $[-u,u]$ is a norm on E_u and the completion $\tilde E_u := (E_u, p_u)^\sim$ is an AM-space with unit u. Since $\tilde E_u \cong C(K_u)$ by Theorem II.1.3, E_u can be identified with a dense Riesz subspace of $C(K_u)$ containing the constant functions; similarly, for $v\in F_+$ the space F_v can be identified with a like subspace of $\tilde F_v \cong C(K_v)$. By definition, G is called *relatively uniformly complete* if each G_w ($w \in G_+$) is already complete under p_w (hence a Banach lattice). As for many constructions in Banach lattice theory, the construction of an appropriate tensor product of Archimedean Riesz spaces and its universal mapping properties are in an essential way based on principal ideals and their relations to AM-spaces (Part II, §1).

THEOREM 2.2 (Fremlin): *Let E, F be Archimedean Riesz spaces. There exist an (essentially unique) Archimedean Riesz space $E \overline{\otimes} F$ and a Riesz bimorphism $\sigma: E \times F \to E \overline{\otimes} F$, such that σ induces an injection of $E \otimes F$ in $E \overline{\otimes} F$ and the pair $(E \overline{\otimes} F, \sigma)$ possesses the following universal mapping property*:

Given any Archimedean Riesz space G and any Riesz bimorphism, $\psi: E \times F \to G$, there exists a unique Riesz homomorphism, $\tau: E \overline{\otimes} F \to G$ for which $\psi = \tau \circ \sigma$.

In addition, $E \overline{\otimes} F$ has these properties:

(i) *For each $w \in E \overline{\otimes} F$ there exist elements $u \in E_+$, $v \in F_+$ such that $|w - z| \leq \delta(u \otimes v)$ for preassigned $\delta > 0$ and suitable $z \in E \otimes F$.*

(ii) *For each $w > 0$ in $E \overline{\otimes} F$, there exist elements $x \in E_+$, $y \in F_+$ such that $0 < x \otimes y \leq w$.*

(iii) *If G is an Archimedean Riesz space which is relatively uniformly complete, then for each positive bilinear map $\phi: E \times F \to G$ there exists a unique positive linear map $\hat{\phi}: E \overline{\otimes} F \to G$ such that $\phi = \hat{\phi} \circ \sigma$.*

Proof: We divide the proof into three main steps.

1. *Construction of $E \overline{\otimes} F$.* Consider any pair $(u, v) \in E_+ \times F_+$. Identifying E_u and F_v (as above) with dense Riesz subspaces of $\tilde{E}_u \cong C(K_u)$, $\tilde{F}_v \cong C(K_v)$, respectively, the tensor product $E_u \otimes F_v$ becomes a vector subspace of $C(K_u) \otimes C(K_v)$ which, in turn, can be viewed as a subspace of $C(K_u \times K_v)$ (namely, the subspace of all functions $h(s, t) = \sum f_i(s) g_i(t)$ where $f_i \in C(K_u)$, $g_i \in C(K_v)$). We now define $E_u \overline{\otimes} F_v$ as the Riesz subspace of $C(K_u \times K_v)$ generated by $E_u \otimes F_v$. Our aim is to show that whenever $0 \leq u \leq u_1$ in E and $0 \leq v \leq v_1$ in F, then there exists a natural, injective Riesz homomorphism $\phi(u, u_1; v, v_1)$ of $E_u \overline{\otimes} F_v$ into $E_{u_1} \overline{\otimes} F_{v_1}$. To this end, consider the principal ideal I of $C(K_{u_1} \times K_{v_1})$ generated by $\{u \otimes v\}$.

Then it is easy to see that $C(K_u \times K_v)$ is Riesz and norm isomorphic to a closed Riesz subspace of I, I being considered (cf. II.1.4) an AM-space with unit ball the order interval $[-u \otimes v, u \otimes v]$ of $C(K_{u_1} \times K_{v_1})$. (Indeed $C(K_u \times K_v)$ is a closed subalgebra of I containing the unit $u \otimes v$ of I; in general, it is not all of I.) Therefore, $E_u \overline{\otimes} F_v$ can be viewed as the Riesz subspace of $C(K_{u_1} \times K_{v_1})$ generated by $E_u \otimes F_v$; since $E_u \otimes F_v \subset E_{u_1} \otimes F_{v_1}$, $E_u \overline{\otimes} F_v$ can naturally be identified with a Riesz subspace of $E_{u_1} \overline{\otimes} F_{v_1}$. This identification defines the desired homomorphism $\phi(u, u_1; v, v_1)$.

It is clear that these homomorphisms compose as they should; that is, if $0 \leq u \leq u_1 \leq u_2$ in E and $0 \leq v \leq v_1 \leq v_2$ in F, then $\phi(u, u_2; v, v_2) = \phi(u_1, u_2; v_1, v_2) \circ \phi(u, u_1; v, v_1)$. Thus the family $\{E_u \overline{\otimes} F_v : (u, v) \in E_+ \times F_+\}$ forms an inductive system with respect to the natural ordering of $E_+ \times F_+$ and the maps ϕ; we define $E \overline{\otimes} F := \varinjlim E_u \overline{\otimes} F_v$ as the corresponding inductive limit. The space $E \overline{\otimes} F$ can also be viewed as the union of the Riesz spaces $E_u \overline{\otimes} F_v$ with identification of elements related by the maps ϕ. In this sense each element $w \in E \overline{\otimes} F$ is contained in some $E_u \overline{\otimes} F_v$; we define $w \geq 0$ if w is positive in $E_u \overline{\otimes} F_v$. This definition is unambiguous, since each map $\phi(u, u_1; v, v_1)$ is an injective Riesz homomorphism; it follows that $E \overline{\otimes} F$ is an Archimedean Riesz space because each $E_u \overline{\otimes} F_v$ is such a space (as a Riesz subspace of $C(K_u \times K_v)$). From the construction it also follows that there is a natural (linear) injection $\rho: E \otimes F \to E \overline{\otimes} F$ and we observe that, if χ denotes the canonical bilinear map $E \times F \to E \otimes F$, the composition $\sigma := \rho \circ \chi$ is a Riesz bimorphism.

2. *Universal Mapping Property and Unicity of $E \overline{\otimes} F$.* As before, we identify each $E_u \overline{\otimes} F_v$ with its canonical image in $E \overline{\otimes} F$ so that $E \overline{\otimes} F$ is the union of the upward directed family $\{E_u \overline{\otimes} F_v : (u, v) \in E_+ \times F_+\}$. If $\psi: E \times F \to G$ is a given Riesz bimorphism, then so is its restriction $\psi_{u,v}$ to $E_u \times F_v$; hence, $\psi_{u,v}$ can be considered a Riesz bimorphism into $\tilde{G}_w$, where $w := \psi(u, v)$. Since $\tilde{G}_w \cong C(Z_w)$ by a now-familiar argument, Proposition 2.1 implies that $\psi_{u,v}$ has a continuous extension factoring through $C(K_u \times K_v)$: $\tilde{E}_u \times \tilde{F}_v \to C(K_u \times K_v) \to C(Z_w)$, where the second map is a Riesz homomorphism $\tilde{\tau}_{u,v}$. Since G is a Riesz subspace of $C(Z_w)$, $\tilde{\tau}_{u,v}^{-1}(G)$ is a Riesz subspace of $C(K_u \times K_v)$ which contains $E_u \otimes F_v$ and hence $E_u \overline{\otimes} F_v$; we denote the restriction of $\tilde{\tau}_{u,v}$ to $E_u \overline{\otimes} F_v$ by $\tau_{u,v}$. It is readily

verified that the Riesz homomorphisms $\tau_{u,v}$ $((u,v)\in E_+\times F_+)$ are consistent on $E\overline{\otimes}F$; that is, we have $\tau_{u,v}=\tau_{u_1,v_1}\circ\phi(u,u_1;v,v_1)$ whenever $0\leqslant u\leqslant u_1$ and $0\leqslant v\leqslant v_1$. Therefore, the family $\{\tau_{u,v}\}$ defines a Riesz homomorphism $\tau:E\overline{\otimes}F\to G$ which satisfies $\psi=\tau\circ\sigma$.

It is important to note that τ is unique. In fact, each $\tau_{u,v}$ is unique because it is continuous on $E_u\overline{\otimes}F_v\subset C(K_u\times K_v)$ into $\tilde{G}_w$, and completely determined by ψ on the dense vector subspace $E_u\otimes F_v$ of $C(K_u\times K_v)$. The uniqueness of τ implies the uniqueness of the pair $(E\overline{\otimes}F,\sigma)$ to within Riesz isomorphism: If (H,σ') is another pair with the desired properties, we obtain a commuting diagram

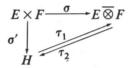

where τ_1, τ_2 are Riesz homomorphisms. Since the identity maps of H and $E\overline{\otimes}F$ are Riesz homomorphisms, the unicity of τ required in the statement of Theorem 2.2 implies $\tau_2\circ\tau_1=1_H$ and $\tau_1\circ\tau_2=1_{E\overline{\otimes}F}$; thus, τ_1 is a Riesz isomorphism of H onto $E\overline{\otimes}F$, and $\sigma=\tau_1\circ\sigma'$.

3. *Additional Properties of $E\overline{\otimes}F$.* (i) Let $w\in E\overline{\otimes}F$. There exist elements $u\in E_+$, $v\in F_+$ such that $w\in E_u\overline{\otimes}F_v$; but $E_u\otimes F_v$ is dense in $C(K_u\times K_v)$ so for each $\delta>0$, there exists $z\in E_u\otimes F_v$ such that $\|w-z\|\leqslant\delta$. In terms of the lattice structure of $C(K_u\times K_v)$ this means that $|w-z|\leqslant\delta(u\otimes v)$ as claimed.

(ii) Again, identifying $w>0$ in $E\overline{\otimes}F$ with an element of some $E_u\overline{\otimes}F_v$ and hence with a continuous function on $K_u\times K_v$, there exist elements $\tilde{x}\in C(K_u)_+$ and $\tilde{y}\in C(K_v)_+$ satisfying $0<\tilde{x}\otimes\tilde{y}\leqslant w$. Since E_u is dense in $C(K_u)$, there exists $\bar{x}\in E_u$ satisfying $|\tilde{x}-\bar{x}|\leqslant\epsilon u$ for preassigned $\epsilon>0$ and, similarly, $|\tilde{y}-\bar{y}|\leqslant\epsilon v$ for suitable $\bar{y}\in F_v$. We let $x:=(\bar{x}-\epsilon u)^+\in E_u$ and $y:=(\bar{y}-\epsilon v)^+\in F_v$, choosing ϵ sufficiently small to ensure $(\tilde{x}-2\epsilon u)^+>0$, $(\tilde{y}-2\epsilon v)^+>0$. Then $0<x\leqslant\tilde{x}$, $0<y\leqslant\tilde{y}$ and so $0<x\otimes y\leqslant w$.

(iii) If $\phi:E\times F\to G$ is a positive bilinear map and G is relatively uniformly complete, then for each $(u,v)\in E_+\times F_+$ the restriction

$\phi_{u,v}$ of ϕ to $E_u \times F_v$ can be factored, by Proposition 2.1, in the manner $E_u \times F_v \to C(K_u \times K_v) \to G_z$ where $z := \phi(u,v)$. Thus $\phi_{u,v}$ can be factored $E_u \times F_v \to E_u \overline{\otimes} F_v \to G$ where the second map is a positive linear mapping $\hat{\phi}_{u,v} : E_u \overline{\otimes} F_v \to G$. It follows now (as in step 2 above) that the family $\{\hat{\phi}_{u,v} : (u,v) \in E_+ \times F_+\}$ determines a unique positive linear map $\hat{\phi} : E \overline{\otimes} F \to G$ satisfying $\phi = \hat{\phi} \circ \sigma$.

This completes the proof of Theorem 2.2. □

COROLLARY: *Let E, F, G be Archimedean Riesz spaces and let $\psi : E \times F \to G$ be a Riesz bimorphism. The associated Riesz homomorphism $\tau : E \overline{\otimes} F \to G$ is injective if and only if $\psi(x, y) > 0$ whenever $x > 0$ and $y > 0$.*

The proof follows at once from (ii) of Theorem 2.2 and can be left to the reader.

3. The *p*-Tensor Product of Banach Lattices.

Let E, F denote Riesz spaces. In accordance with established usage for linear operators (Part I, §5), a bilinear form $\phi : E \times F \to \mathbf{R}$ is called *regular* if $\phi = \phi_1 - \phi_2$ for suitable positive bilinear forms ϕ_1, ϕ_2 (cf. §2). We recall that for Banach lattices E, F, the space $\mathcal{L}^r(E, F')$ is a Banach lattice (I.5.3).

PROPOSITION 3.1: *Let E, F denote Banach lattices. The identity*

$$\phi(x, y) = \langle T_\phi x, y \rangle \qquad (x \in E, y \in F)$$

defines a Riesz isomorphism $\phi \mapsto T_\phi$ of the space $\mathcal{B}^r(E, F)$ of all regular bilinear forms on $E \times F$, ordered by the relation "$\phi \leq \psi$ iff $\psi - \phi$ is positive," onto the space $\mathcal{L}^r(E, F')$. In particular, $\mathcal{B}^r(E, F)$ is an order complete Banach lattice under this ordering and the norm $\phi \mapsto \|\phi\|_r := \sup\{|\phi|(x, y) : \|x\| \leq 1, \|y\| \leq 1\}$.

Proof: The proof consists of essentially two observations. First, a positive bilinear form on $E \times F$ determines a positive linear operator of E into F^*, and conversely; and $F^* = F'$ by Theorem I.5.1. Second, the norm of $\mathcal{B}^r(E, F)$ defined above thus corresponds to the *r*-norm and so by Proposition I.5.3, $\mathcal{B}^r(E, F)$ is an order complete Banach lattice. □

Since by Theorem I.5.1, each positive linear operator $E \to F^*$ has range in F' and is continuous, we obtain the following corollary.

COROLLARY: *If E and F are Banach lattices, every regular bilinear form on $E \times F$ is continuous.*

In §1 we have shown that for Banach spaces G, H, the operator space $\mathcal{L}(G, H')$ is actually a dual Banach space; namely, the dual of $G \otimes_\pi H$. Our aim is now, via the tensor product constructed in §2, to identify $\mathcal{L}^r(E, F')$ as a dual Banach lattice with respect to a suitable (completed) tensor product of E and F. We recall that a norm on $E \otimes F$ is called a *cross-norm* if $\|x \otimes y\| = \|x\| \|y\|$ for all $x \in E$ and $y \in F$, the norms on the right-hand side denoting the given norm of E and F, respectively; a cross-norm is called *symmetric* if the natural map $E \otimes F \to F \otimes E$ is an isometry. (Not all useful cross-norms are symmetric; cf. [16, Chapter IV, §7].) If, as in our present situation, E and F are Banach lattices, the convex hull of $E_+ \otimes F_+$ (i.e., the set of all linear combinations $\sum \alpha_i x_i \otimes y_i$ where $\alpha_i \geq 0$, $x_i \in E_+$, $y_i \in F_+$) is a convex cone called the *projective cone* C_p of $E \otimes F$. (It follows from Theorem 2.2 that C_p is contained in the positive cone of $E \overline{\otimes} F$ and hence defines a vector ordering of $E \otimes F$.)

THEOREM 3.2: *Let E, F denote Banach lattices. There exists a symmetric cross-norm p on $E \otimes F$ having these properties*:
  (i) *The completion $E \tilde{\otimes}_p F$ is a Banach lattice under the ordering whose positive cone is the closure of C_p in $E \tilde{\otimes}_p F$.*
  (ii) *The dual of $E \tilde{\otimes}_p F$ is Riesz and norm isomorphic to $\mathcal{L}^r(E, F')$.*
  (iii) *p is the greatest cross-norm on $E \otimes F$ having property* (i).

Proof: (i) Let G denote the dual Banach lattice $\mathcal{L}^r(E, F')'$ and consider the bilinear map $\psi \colon E \times F \to G$ given by $\psi(x, y)(T) := \langle Tx, y \rangle$. A moment's reflection shows that, by virtue of Proposition I.3.3, Corollary, ψ is a Riesz bimorphism. Moreover, $\psi(x, y) > 0$ whenever $x > 0$ and $y > 0$, so Theorem 2.2 and its corollary show that $E \overline{\otimes} F$ can be identified with a Riesz subspace of G. The norm closure of $E\overline{\otimes}F$ in G is, therefore, a Banach lattice H; it follows from Theorem 2.2, Property (i), that $E\otimes F$ is dense in H so that we can justly denote H by $E\tilde{\otimes}_p F$, using the letter p for the Riesz norm on H induced by G. It is readily verified that p is a symmetric cross-norm on $E\otimes F$, and it will be shown at once that C_p is dense in the positive cone of $E\tilde{\otimes}_p F$.

(ii) In view of Proposition 3.1 and the definition of p we find

$$p(u)=\sup\{\hat{\phi}(|u|):\phi\in\mathcal{B}^r(E,F),\|\phi\|_r\leq 1\} \qquad (1)$$

for $u\in E\overline{\otimes}F$ where, as before, $\hat{\phi}$ denotes the linear form on $E\overline{\otimes}F$ defined by ϕ. In particular, the p-unit ball of $E\overline{\otimes}F$ contains the set

$$V:=\{u\in E\overline{\otimes}F:|u|\leq \Sigma x_i\otimes y_i, x_i\in E_+, \qquad (2)$$
$$y_i\in F_+, \Sigma\|x_i\|\|y_i\|\leq 1\},$$

because for each $\phi\geq 0$, $\|\phi\|\leq 1$, and $u\in V$ we have $\hat{\phi}(|u|)\leq \Sigma\phi(x_i,y_i)\leq 1$. On the other hand, if f is a linear form on $E\overline{\otimes}F$ bounded by 1 on V, then from Part I, §3, Formula (1) it follows that $|f|$ exists and $|f|(u)\leq 1$ on V. Therefore, the polar V° of V in $\mathcal{B}^r(E,F)\cong\mathcal{L}^r(E,F')$ is contained in (and by definition contains) the unit ball of $\mathcal{L}^r(E,F')$. Hence by the bipolar theorem, V is dense in the p-unit ball of $E\overline{\otimes}F$ and thus of $E\tilde{\otimes}_p F$, and the dual $(E\tilde{\otimes}_p F)'$ is Riesz and norm isomorphic to $\mathcal{L}^r(E,F')$, as asserted.

It is now easy to see that C_p is dense in $(E\tilde{\otimes}_p F)_+$. In fact, the dual cone of C_p in $\mathcal{L}^r(E,F')$ is the set of all operators T for which $\langle Tx,y\rangle\geq 0$ whenever $x\in E_+$ and $y\in F_+$, which is the positive cone of $\mathcal{L}^r(E,F')$; again by the bipolar theorem, $\overline{C}_p=(E\tilde{\otimes}_p F)_+$.

(iii) Suppose q is a cross-norm on $E\otimes F$ such that $E\tilde{\otimes}_q F$ is a Banach lattice under the ordering defined by the q-closure of C_p. If $u\in V\cap(E\otimes F)$, by (2) we have $\pm u\leq \Sigma x_i\otimes y_i$ for suitable elements $x_i\in E_+$, $y_i\in F_+$ satisfying $\Sigma\|x_i\|\|y_i\|\leq 1$; here $\leq$ denotes the ordering defind by the cone $(E\tilde{\otimes}_p F)_+$ whose elements v are precisely those satisfying $\phi(v)\geq 0$ for all $\phi\in\mathcal{B}^r(E,F)_+$. Since the positive cone of $(E\tilde{\otimes}_q F)'$ is contained in $\mathcal{B}^r(E,F)_+$, we have $\pm u\leq \Sigma(x_i\otimes y_i)$ in $E\tilde{\otimes}_q F$ as well and it follows that $q(u)\leq$

$\Sigma q(x_i \otimes y_i) = \Sigma \|x_i\| \|y_i\| \leq 1$ for all $u \in V \cap (E \otimes F)$; thus $q(u) \leq p(u)$ for all $u \in E \otimes F$ as claimed. □

By the preceding proof, the set V (see (2)) is dense in the unit ball of $E \tilde{\otimes}_p F$ and, hence, (2) yields the following explicit characterization of the p-norm; here the relation $|u| \leq \Sigma x_i \otimes y_i$ (or equivalently, $\pm u \leq \Sigma x_i \otimes y_i$) can be interpreted as meaning $\pm \phi(u) \leq \Sigma \phi(x_i, y_i)$ for each positive bilinear form ϕ on $E \times F$.

COROLLARY: *The p-norm of an element $u \in E \otimes F$ is given by*

$$p(u) = \inf \left\{ \sum_{i=1}^{n} \|x_i\| \|y_i\| : x_i \in E_+, y_i \in F_+, n \in \mathbb{N}, \pm u \leq \sum_{i=1}^{n} x_i \otimes y_i \right\}.$$

If E is any Banach lattice and if F is an AL-space, then by Proposition I.5.4 we have $\mathcal{L}^r(E, F') = \mathcal{L}(E, F')$ and thus $\mathcal{B}^r(E, F) = \mathcal{B}(E, F)$; therefore, the p-norm and the π-norm on $E \otimes F$ yield the same dual. It follows from standard arguments that in this case, we have $E \otimes_\pi F = E \otimes_p F$.

In conclusion, by the procedure adopted in the proof of (3.2), all cross norms on $E \otimes F$ can be found that satisfy (i) of the theorem and for which the map $(x, y) \mapsto x \otimes y$ is a Riesz bimorphism. (For examples with interesting applications, see [16, Chapter IV, §§7–8].) A detailed study has been made by H. P. Lotz ("Tensor Products and Lattice Ideals of Operators," preprint, 1974).

4. Extension of Positive Operators. We have now gathered enough information to tackle the problem touched upon at the beginning of §2. We consider the category of all Banach lattices, denoted by **BL**, with morphisms the positive contractions. An object F in **BL** is *injective* if for every Banach lattice E and Banach sublattice E_0 of E, each morphism $T_0: E_0 \to F$ has an extension $T: E \to F$ which is a morphism[†] of **BL**.

[†]It can be shown that the injective objects are the same if **BL** is understood to be the category of all Banach lattices with morphisms the regular operators with r-norm ≤ 1.

PROPOSITION 4.1: *The following assertions on a Banach lattice F are equivalent*:
(a) *The dual F' is injective in* **BL**.
(b) *For every Banach lattice E and Banach sublattice $E_0 \subset E$, the natural map $E_0 \otimes_p F \to E \otimes_p F$ is an isometry.*

Proof: If E_0 is a Banach sublattice of E, the map $(x, y) \mapsto x \otimes y$ is a Riesz bimorphism (§2) of $E_0 \times F$ into $E \tilde{\otimes}_p F$ and hence by Theorem 2.2, $E_0 \overline{\otimes} F$ can be considered a Riesz subspace of $E \tilde{\otimes}_p F$. Now, in view of Theorem 3.2 (ii), the implication (a)$\Rightarrow$(b) results from Part I, §3, Formula (3); the converse (b)$\Rightarrow$(a) follows from Theorem I.3.4. □

THEOREM 4.2: *Every Banach lattice $C(K)$, K compact Stonian, and every* AL-*space is injective in* **BL**.

Proof: For both types of Banach lattices, there exists a positive contractive projection of the bidual onto the given space; for spaces $C(K)$ this results from Gleason's theorem (1.3) while for AL-spaces this results from Part I, Theorem 4.4 and Proposition 4.5 (an AL-space being a closed ideal, hence a projection band, in its bidual). Thus we can restrict attention to the case of a bidual AM-space or AL-space F', F being a dual AL- or AM-space, respectively. If F is an AL-space, then by the remark immediately following the corollary of Theorem 3.2, we have $E_0 \otimes_p F = E_0 \otimes_\pi F$ and $E \otimes_p F = E \otimes_\pi F$; hence by the discussion following Proposition 1.2, $E_0 \otimes_p F \to E \otimes_p F$ is an isometry as required by Proposition 4.1.

The case where F is a dual AM-space (hence where $F = C(K)$ with K compact hyperstonian) is settled as follows. If F_0 denotes the Riesz subspace of all continuous functions on K with finite range, then, since F_0 is dense in F, it suffices to show that $E_0 \otimes_p F_0 \to E \otimes_p F_0$ is an isometry. If $u \in E_0 \otimes F_0$, we have to show that its norm $p_0(u)$ in $E_0 \otimes_p F_0$ equals its norm $p(u)$ in $E \otimes_p F_0$. From the Corollary of Theorem 3.2 it is clear that $p(u) \leq p_0(u)$. Conversely, letting $u = \sum x_i \otimes f_i$ where the elements $f_i \in F_0$ can be assumed to be orthogonal, there exist finite sequences (z_j) in E_+

and (g_j) in $(F_0)_+$ such that $|u| \leq \Sigma z_j \otimes g_j$ and $\Sigma \|z_j\| \|g_j\| < p(u) + \epsilon$ ($\epsilon > 0$ being preassigned). Evidently we can assume here that $\|g_j\| = 1$ for all j and $f_i \geq 0$, $\|f_i\| = 1$ for all i. Now if we let $x := \sup_i x_i$ in E_0, we still have $x \leq \Sigma_j z_j$ and hence, keeping in mind that $|u| \leq x \otimes e$ (e the constant-one function on K), we obtain $p_0(u) \leq \|x\| \leq \Sigma \|z_j\| < p(u) + \epsilon$. Therefore, $p_0(u) = p(u)$ and the imbedding $E_0 \otimes_p F_0 \to E \otimes_p F_0$ is indeed an isometry. □

To the preceding proof the objection can be raised that injectivity of $C(K)$ (K Stonian) in **BL** can be derived directly from Gleason's theorem (1.3). However, our objective is to exhibit the role of the p-tensor product in this context; this proves indeed most fruitful in the surprisingly short and elegant proof (due to D. I. Cartwright) of the following theorem.

THEOREM 4.3: *Let F, G denote Banach lattices. If F' and G are injective in* **BL**, *then so is the Banach lattice* $\mathcal{L}^r(F, G)$.

Proof: We suppose F', G injective and consider G as a Banach sublattice of its bidual G''; this implies the existence of a positive contractive projection $P: G'' \to G$. Hence G is order complete (in fact, if A is directed ($\leq$) and order bounded in G and if $x'' := \sup A$ in G'', then $Px'' = \sup A$ in G), and so $\mathcal{L}^r(F, G)$ is well defined and can be considered a Banach sublattice of $\mathcal{L}^r(F, G'')$. Moreover, the map $T \mapsto PT$ defines a positive contractive projection of $\mathcal{L}^r(F, G'')$ onto $\mathcal{L}^r(F, G)$; therefore, to prove the theorem it suffices to show that $\mathcal{L}^r(F, G'')$ is injective in **BL**.

Now by Theorem 3.2, $\mathcal{L}^r(F, G'')$ is the dual of $H := F \tilde{\otimes}_p G'$ and by Proposition 4.1 we must show that $E_0 \tilde{\otimes}_p H \to E \tilde{\otimes}_p H$ is an isometry for all Banach lattices E and sublattices E_0. By the injectivity of F', $E_0 \tilde{\otimes}_p F \to E \tilde{\otimes}_p F$ is an isometry and by the injectivity of G, $(E_0 \tilde{\otimes}_p F) \tilde{\otimes}_p G \to (E \tilde{\otimes}_p F) \tilde{\otimes}_p G$ is an isometry because, as we just saw, $E_0 \tilde{\otimes}_p F$ can be identified with a Banach sublattice of $E \tilde{\otimes}_p F$. Now it is an important but elementary fact that the p-tensor product is associative, and thus we find that $E_0 \tilde{\otimes}_p (F \tilde{\otimes}_p G')$ imbeds isometrically as a Riesz subspace of $E \tilde{\otimes}_p (F \tilde{\otimes}_p G')$; again by Proposition 4.1 it follows that $(F \tilde{\otimes}_p G')' = \mathcal{L}^r(F, G'')$ is injective. □

This result brings out a host of injective objects in **BL** which are neither AM- nor AL-spaces; in particular, whenever $F = G$ is an AL-space or order complete AM-space with unit, the operator algebra $\mathcal{L}^r(F)$ ($\cong \mathcal{L}(F)$) is injective in **BL**. A complete characterization of injective objects in **BL** has recently been given by R. Haydon (Math. Z., **156** (1977), 19–47).

PART IV. CYCLIC BANACH SPACES

1. Bounded Boolean Algebras of Projections. We suppose the reader to be familiar with the definition of a Boolean algebra, recalling its lattice theoretic version: A *Boolean algebra* is a distributive, complemented lattice $(A, \vee, \wedge)$ with greatest element 1 and least element 0. The simplest example of a Boolean algebra is the power set 2^X of any set X, where $\vee$ and $\wedge$ are set theoretic union and intersection, respectively. If X is a topological space, a slightly more sophisticated example is the set $\mathbf{Q}(X)$ of all open-and-closed subsets of X (which is a Boolean subalgebra of 2^X); the importance of this example stems from the famous and well-known theorem of M. H. Stone to the effect that every Boolean algebra A is isomorphic to $\mathbf{Q}(K)$ for a unique, totally disconnected compact (Hausdorff) space K commonly referred to as the *Stone space* of A (cf. [12, I.8.7] or [16, Chapter II, Exercise 1]).

Let G denote a real Banach space.[†] By a *Boolean algebra of projections* (B.a.p.) on G we will understand a bounded set **P** of (continuous) projection operators on G such that:

P *is a Boolean algebra under the operations* $P \vee Q := P + Q - PQ$ *and* $P \wedge Q := PQ$, *with* 0 *the zero operator and* 1 *the identity operator on* G.

It follows trivially from this definition that every B.a.p. **P** on G is a commutative subset of $\mathcal{L}(G)$. If there exists a vector $x_0 \in G$ such that G equals the closed linear hull of $\{Px_0 : P \in \mathbf{P}\}$, then x_0

[†]The restriction to real scalars is merely a convenience; with a suitable notion of complex Banach lattice [16, Chapter II, §11], all results of this chapter remain valid for B.a.p.'s on a complex Banach space G.

is called a *cyclic vector* and G is called a *cyclic Banach space* (with respect to **P** and x_0). We consider some examples:

1. For any p satisfying $1 \leq p < +\infty$, let G denote the sequence space l^p and let x_0 denote the sequence $(2^{-n})_{n \in \mathbf{N}}$. For any subset A of **N** denote by $P_A \in \mathcal{L}(l^p)$ the projection $x \mapsto \sum_{n \in A} \xi_n e_n$ (where $x = (\xi_n)$ and e_n denotes the nth unit vector). Let $\mathbf{A}_0$ be the Boolean algebra of all subsets of **N** which are either finite or have finite complement, and let $\mathbf{A} = 2^{\mathbf{N}}$. Then $\mathbf{P}_0 := \{P_A : A \in \mathbf{A}_0\}$ and $\mathbf{P} := \{P_A : A \in \mathbf{A}\}$ are B.a.p.'s on l^p; moreover, with respect to x_0 and either of $\mathbf{P}_0$, **P** the space l^p is a cyclic Banach space. For $p=2$, each P_A ($A \subset \mathbf{N}$) is Hermitian (orthogonal).

2. Let G be the Banach space of all bounded Borel functions on $[0,1]$, endowed with the supremum norm. For any Borel set $B \subset [0,1]$ let P_B denote the projection on G defined by $P_B f := \chi_B f$ (pointwise multiplication). The family **P** of all such projections is a B.a.p., and G is cyclic with respect to **P** and the constant-one function e.

3. Let E denote a Banach lattice with order continuous norm (Part I, §4) and topological unit x_0 (Part II, §4). If **P** denotes the Boolean algebra of all band projections of E (cf. Part I, Propositions 2.3 and 2.4), it is not hard to prove [16, II.2.13] that the set $\{Px_0 : P \in \mathbf{P}\}$ is the extreme boundary of the order interval $[0, x_0]$; since $[0, x_0]$ is weakly compact by Theorem I.4.4, the Krein-Milman theorem shows $[0, x_0]$ to be the convex closure of $\{Px_0 : P \in \mathbf{P}\}$. Since $[0, x_0]$ is total in E, by definition of a topological unit, the Banach space E is cyclic with respect to x_0 and **P**. (Note that Example 1, with respect to the B.a.p. **P** defined there, is a special case of Example 3; however, Example 2 is not.)

We next investigate the norm closed algebra $A(\mathbf{P}) \subset \mathcal{L}(G)$ generated by **P**.

PROPOSITION 1.1: *The Banach algebra $A(\mathbf{P})$ is isomorphic (not necessarily isometric) with the Banach algebra $C(K)$, where K denotes the Stone space of* **P**.

Proof: Let S denote the subalgebra of $C(K)$ consisting of all functions f with finite range; each $f \in S$ is of the form $\sum_{i=1}^{n} \alpha_i \chi_i$ where χ_i is the characteristic function of an open-and-closed subset U_i of K, and the sets U_i can be assumed to be disjoint. Since K is totally disconnected (i.e., since $\mathbf{Q}(K)$ is a base of K), S is dense in $C(K)$. (This can be easily verified directly, without appeal to the Stone-Weierstrass theorem.) Let $\tilde{m}$ denote the Stone isomorphism of $\mathbf{Q}(K)$ onto $\mathbf{P}$; we define a mapping $m: S \to \mathcal{L}(G)$ by letting $m(f) := \sum_{i=1}^{n} \alpha_i \tilde{m}(U_i)$. Now let M denote a bound for the norms of all $P \in \mathbf{P}$ and let $x \in G$, $x' \in G'$ be fixed. Supposing the sets $U_i \in \mathbf{Q}(K)$ to be pairwise disjoint, a set U_i is called of the first kind if $\langle \tilde{m}(U_i)x, x'\rangle \geq 0$; otherwise it is of the second kind. Denoting by U (respectively, V) union of the sets of the first (respectively, second) kind and denoting the corresponding summations by $'$ (respectively, $''$), we obtain

$$|\langle m(f)x, x'\rangle| \leq \sum{}'\langle|\alpha_i|\langle \tilde{m}(U_i)x, x'\rangle - \sum{}''|\alpha_i|\langle \tilde{m}(U_i)x, x'\rangle$$

$$\leq \sup_i |\alpha_i|(\|\tilde{m}(U)\| + \|\tilde{m}(V)\|)\|x\|\,\|x'\|$$

$$\leq 2M\|f\|\,\|x\|\,\|x'\|.$$

Thus $\|m(f)\| \leq 2M\|f\|$ for all $f \in S$. On the other hand, if $f = \sum \alpha_i \chi_i \neq 0$ with disjoint χ_i, then $\|f\| = |\alpha_j|$ for a suitable j; considering a vector $x \in \tilde{m}(U_j)G$, $\|x\| = 1$, we find $m(f)x = \alpha_j x$ and hence $\|m(f)x\| = |\alpha_j| = \|f\|$. Thus $m: S \to \mathcal{L}(G)$ is an algebra homomorphism satisfying $\|f\| \leq \|m(f)\| \leq 2M\|f\|$; its continuous extension to $C(K)$, which we again denote by m, is an isomorphism of Banach algebras $C(K) \to A(\mathbf{P})$. □

PROPOSITION 1.2: *The norm on G defined by*

$$\|x\|_1 := \sup_{\|f\| \leq 1} \|m(f)x\| \qquad (x \in G) \tag{*}$$

is equivalent to the given norm of G. With respect to the corresponding operator norm on $\mathcal{L}(G)$, the isomorphism $m: C(K) \to A(\mathbf{P})$ is an isometry and induces an isomorphism of the group of all unimodular functions in $C(K)$ onto a group of contractions in $\mathcal{L}(G)$.

Proof: If now U denotes the unit ball of $C(K)$ and if $f \in U$, then for any $x \in G$ we have

$$\|m(f)x\|_1 = \sup_{g \in U} \|m(g)m(f)x\| = \sup_{g \in U} \|m(gf)x\| \leq \|x\|_1,$$

because m is multiplicative; thus $\|m(f)\| \leq 1$ in the operator norm associated with the norm (*) on G. On the other hand, let $f \in S$ and $\|f\| = 1$; as in the proof of (1.1) there exists $x \in G$, $\|x\| = 1$, satisfying $m(f)x = x$, and hence we have $\|m(f)\| \geq 1$. Thus m is an isometry on S and, therefore, on $C(K)$. Finally, if $f \in C(K)$ is a unimodular function ($|f(t)| = 1$ for all $t \in K$), then $fU = U$; therefore, $\|m(f)x\|_1 = \|x\|_1$ for each $x \in G$ and $m(f)$ is an isometry for the norm (*) on G. □

THEOREM 1.3: *Let G be a cyclic Banach space with respect to $\mathbf{P}$ and x_0. There exists an ordering of G such that under the equivalent norm (*) of (1.2), G becomes a Banach lattice with topological unit x_0. Moreover, $\mathbf{P}$ acts on E as a Boolean algebra of band projections.*

Proof: Consider the mapping $m_0: C(K) \to G$ defined by $m_0(f) := m(f)x_0$; by definition of a cyclic vector in G, $m_0(C(K))$ is a dense linear subspace of G. It is clear that the subalgebra S of all functions with finite range in $C(K)$ is also a Riesz subspace of $C(K)$; if $h, g \in S$ satisfy $|h| \leq |g|$, it is readily seen that there exists a function $p \in S$, $\|p\| \leq 1$, such that $h = pg$. Therefore,

$$\|m(h)x_0\|_1 = \|m(p)m(g)x_0\|_1 \leq \|m(g)x_0\|_1 \tag{1}$$

by virtue of Proposition 1.2; by continuity it follows that (1) is equally valid whenever $h, g \in C(K)$ and $|h| \leq |g|$. Moreover, (1) implies that m_0 is injective; in fact if $f \in C(K)$, $f \neq 0$, there exists $\alpha > 0$ and a non-void set $U \in \mathbf{Q}(K)$ such that $|f| \geq \alpha \chi_U$, whence $\|m(f)x_0\| \geq \alpha \|\tilde{m}(U)x_0\| > 0$.

Thus under m_0 we can identify $C(K)$ with a dense subspace of G, and by (1) the norm (*) of G induces a Riesz norm on $C(K)$. By the discussion following Axiom (NL) (Part I, §1) the lattice operations of $C(K)$ have unique continuous extensions to G and thus G becomes a Banach lattice E. Since $m_0(e) = x_0$, the ideal E_{x_0} contains $C(K)$ and so is dense: x_0 is a topological unit of E. Finally, each $P \in \mathbf{P}$ satisfies $0 \leq Px \leq x$ for all $x \in E_+$ under this ordering; hence by the remark preceding Proposition I.2.3, each $P \in \mathbf{P}$ acts as a band projection on E. □

2. σ-Complete Boolean Algebras of Projections.

Much as it may surprise the reader that every cyclic Banach space G is essentially a Banach lattice on which the given B.a.p. **P** acts as a Boolean algebra of band projections, it can occur (as it does for the B.a.p. $\mathbf{P}_0$ of Example 1 (§1)) that the space G is not very well characterized by **P**. Also, as shown by Example 2 (§1), the Banach lattice E obtained need not have order continuous norm or even be order complete.

It will be shown in this section that the seemingly much more special situation of Example 3 (§1) is always at hand if **P** is required to be σ-complete in the following sense. A B.a.p. **P** on a Banach space G is called *σ-complete* (respectively, *complete*) if for each countable directed ($\leq$) family (P_α) in P (respectively, if for every directed ($\leq$) family (P_α) in P), the upper bound $P := \sup_\alpha P_\alpha$ exists in **P** and if $\lim_\alpha \langle P_\alpha x, x' \rangle = \langle Px, x' \rangle$ for each pair $(x, x') \in G \times G'$. (Veksler [18] and Rall [14] employ the following weaker concept: **P** is called *τ-complete* if for each $x \in G$ there exists a smallest projection $P_x \in \mathbf{P}$ leaving x fixed.)

Suppose now that **P** is a σ-complete B.a.p. on G and, as before, let K denote the Stone space of **P**. The Boolean σ-algebra $\Sigma \subset 2^K$ generated by $\mathbf{Q}(K)$ (which is the Baire field of K, since K is totally disconnected) has the property that, for each $B \in \Sigma$, there exists a (unique) $U \in \mathbf{Q}(K)$ such that the symmetric difference $B \dotplus U$ is in Δ, where Δ denotes the σ-ring generated by all rare closed G_δ-subsets of K; it suffices in fact to see that the family of all subsets $B \subset K$ having this property is a σ-algebra. The mapping $q: \Sigma \to \mathbf{Q}(K)$ defined by $B \mapsto U$ is even a σ-homomorphism (i.e., a homomorphism preserving countable suprema) of Boolean algebras. Again denoting by $\tilde{m}: \mathbf{Q}(K) \to \mathbf{P}$ the Stone isomorphism and selecting an arbitrary pair $(x, x') \in G \times G'$, the composition $\tilde{m} \circ q$ gives us a mapping $B \mapsto \langle \tilde{m}(q(B))x, x' \rangle$ of Σ into **R**; by the continuity property required of **P**, this mapping is just a bounded, signed Baire measure $\mu_{x,x'}$ on K. We note that by this definition, each set in Δ is a null set for every $\mu_{x,x'}$. The multiplicativity of $\tilde{m}$ (that is, the property $\tilde{m}(U \cap V) = \tilde{m}(U)\tilde{m}(V)$ for $U, V \in \mathbf{Q}(K)$), implies the following decisive property of the Baire measures $\mu_{x,x'}$.

LEMMA 2.1: *Let **P** be a σ-complete B.a.p. with bound M in $\mathcal{L}(G)$ and let $x \in G$ be fixed. For each $x' \in G'$ there exists a decomposition*

$x' = x_1' - x_2'$ such that $\|x_i'\| \leq M\|x'\|$ and such that the measures $\mu_{x,x_i'}$ are positive ($i=1,2$).

Proof: Since $\mu_{x,x'}$ is a Baire measure, there exists a partition of K into Baire sets A_+, A_- such that $\mu_{x,x'}$ is positive on (every Baire subset of) A_+ and negative on A_- (Jordan decomposition). Let $x_1' := (\tilde{m} \circ q(A_+))'x'$ and $x_2' := (\tilde{m} \circ q(A_-))'x'$. For $B \in \Sigma$, we have $\mu_{x,x_1'}(B) = \langle \tilde{m} \circ q(B)x, (\tilde{m} \circ q(A_+))'x' \rangle = \langle \tilde{m} \circ q(A_+) \tilde{m} \circ q(B)x, x' \rangle = \langle \tilde{m} \circ q(A_+ \cap B)x, x' \rangle = \mu_{x,x'}(A_+ \cap B) \geq 0$; similarly, $\mu_{x,x_2'}(B) \geq 0$ and $\mu_{x,x'} = \mu_{x,x_1'} - \mu_{x,x_2'}$. Since $\|\tilde{m}(U)\| \leq M$ for all $U \in \mathbf{Q}(K)$, it is clear that $\|x_i'\| \leq M\|x'\|$. □

The following lemma is now the decisive step in proving the announced result. As in §1, we denote by m the algebra homomorphism $C(K) \to \mathcal{L}(G)$ induced by the Stone representation $\tilde{m}$: $\mathbf{Q}(K) \to \mathbf{P}$.

LEMMA 2.2: *Let $\mathbf{P}$ denote a σ-complete B.a.p. on the Banach space G, and let $x_0 \in G$ be fixed. Under the mapping $m_0: f \mapsto m(f)x_0$, the image of the unit ball U of $C(K)$ is closed in G.*

Proof: Let (f_n) denote a sequence in U for which $(m_0(f_n))$ is Cauchy in G; without loss of generality we can assume that $\|m_0(f_{n+1} - f_n)\| \leq M^{-1}2^{-n-1}$ for all n. Also, it suffices to show that $(m_0(f_n))$ converges weakly to $m_0(g)$, where g is a suitable element of U.

Since $\mathbf{P}$ is σ-complete, its Stone space K has the property that the closure of each open F_σ-subset is open (i.e., K is quasi-Stonian), or equivalently, that $C(K)$ is countably order complete. (The proof of this relationship is quite analogous to that of Proposition II.1.5; for details, see [16, Chapter II, 7.7 and 7.8].) The set $B_n := \{s \in K : f_{n+1}(s) - f_n(s) \neq 0\}$ is an open F_σ-set and the function sign $(f_{n+1} - f_n)$ is continuous on B_n; hence this function has a continuous extension h_n with norm ≤ 1 to all of K (cf. II.1.6). Clearly, $|f_{n+1} - f_n| = h_n(f_{n+1} - f_n)$ and from the multiplicativity of m we conclude that $\|m_0(|f_{n+1} - f_n|)\| \leq \|m(h_n)\| \|m_0(f_{n+1} - f_n)\| \leq 2^{-n}$.

We now have to find a function $g \in U$ such that $\lim_n \int (f_n - g) d\mu_{x_0, x'} = 0$ for each $x' \in G'$; by Lemma 2.1 it is enough to consider those $x' \in G'$ for which $\mu_{x, x'} \geq 0$. Since $\int |f_{n+1} - f_n| d\mu_{x_0, x'} \leq \|m_0(|f_{n+1} - f_n|)\| \|x'\| \leq 2^{-n} \|x'\|$, the series

$$\sum_{n=1}^{\infty} \int |f_{n+1} - f_n| d\mu_{x_0, x'}$$

converges. By B. Levi's theorem, $\lim_n f_n(s)$ exists a.e. ($\mu_{x_0, x'}$) in K; wherever it exists, it agrees with $\bar{g}(s) = \limsup_n f_n(s)$, the numerical upper limit $\bar{g}$ of the sequence being a Baire function absolutely bounded by 1. Since $C(K)$ is countably order complete, we can also define the functions $g_n := \sup\{f_\nu; \nu \geq n\}$ and $g := \inf_n g_n$ in $C(K)$; it is seen without difficulty that g differs from $\bar{g}$ only in a meager Baire set $N \subset K$. Thus $N \in \Delta$ and, as noted above, N is a null set for $\mu_{x_0, x'}$. Therefore we have $\lim_n f_n(s) = g(s)$ for $\mu_{x_0, x'}$-almost all $s \in K$ and hence, by the dominated convergence theorem, $\lim_n \int |f_n - g| d\mu_{x_0, x'} = 0$ for each $x' \in G'$ for which $\mu_{x_0, x'} \geq 0$. By (2.1) this means that $\lim_n m_0(f_n) = m_0(g)$ weakly in G and the proof is complete. □

THEOREM 2.3: *Let G be a cyclic Banach space with respect to $x_0 \in G$ and the σ-complete B.a.p. **P**. There exist an ordering and equivalent norm (given by (*) of (1.2)) under which G becomes a Banach lattice E with order continuous norm and topological unit x_0. Moreover, **P** is complete and acts on E as the Boolean algebra of all band projections, and the Stone space K of **P** is homeomorphic with the structure space K_E of E.*

Proof: We only have to supplement the proof of Theorem 1.3. Let E denote the Banach lattice with topological unit constructed in that proof. The injective mapping $m_0: C(K) \to G$ takes $C(K)$ onto a dense subspace of G and, in particular, the unit ball U of $C(K)$ onto a dense subset of the order interval $[-x_0, x_0]$ of E. Since $m_0(U)$ is closed in G by Lemma 2.2, m_0 maps $C(K)$ onto a dense ideal of E and this identifies K as the structure space K_E of E (II.4.3). In turn, this shows **P** to induce the Boolean algebra $\mathbf{P}_E$ of all band projections of E. Indeed, if $P \in \mathbf{P}_E$, then P induces a band projection of the ideal $C(K)$, and $e_1 = Pe$, $e_2 = (1-P)e$ (e the unit if $C(K)$) are orthogonal continuous functions satisfying $e_1 + e_2 = e$. This means that $e_1 = \chi_V$ for some $V \in \mathbf{Q}(K)$, and hence $P = \tilde{m}(V) \in \mathbf{P}$.

To show that E has order continuous norm, we use criterion (f) of Theorem I.4.4. Supposing (x_n) to be an orthogonal sequence in $[0,x] \subset E$, we must show that $\|x_n\| \to 0$. Obviously we can replace x by $x_0 \vee x$ if necessary, so we assume that $x \geqslant x_0$. Then x is a topological unit of E, and by Proposition II.4.1 there exists a Riesz isomorphism $p: E_{x_0} \to E_x$. Since $p \circ m_0$ is a Riesz isomorphism of $C(K)$ onto E_x, x is a cyclic vector of G and we may as well assume that $x = x_0$. (Note that replacing x_0 by x affects neither the order nor the norm of E.) Under this assumption we have $x_n = m_0(f_n)$ ($n \in \mathbf{N}$) for an orthogonal sequence (f_n) in $[0,e] \subset C(K)$. Since $\mathbf{P}$ is σ-complete by hypothesis, K is quasi-Stonian and $C(K)$ order σ-complete. Consider $g_n := \sup_{\nu \geqslant n} f_\nu$ in $C(K)$; (g_n) is decreasing and we have $\inf_n g_n = 0$, which implies that $\lim_n g_n(s) = 0$ outside a meager Baire set $N \subset K$. Thus $\lim_n \langle m_0(g_n), x' \rangle = \lim_n \int g_n \, d\mu_{x_0, x'} = 0$ for each $x' \in E'_+$, where N is a null set for each of the positive Baire measures $\mu_{x, x'}$. By Proposition I.4.1, $\|m_0(g_n)\| \to 0$ and hence, because of $0 \leqslant x_n \leqslant m_0(g_n)$, we have $\|x_n\| \to 0$ in E.

Finally, since E has order continuous norm, by Proposition I.4.5 the closed ideals are identical with the projection bands of E; this implies that $\mathbf{P}_E$, and hence $\mathbf{P}$, is a complete B.a.p. □

In view of Example 3 (§1), it follows that a Banach space is cyclic with respect to some σ-complete B.a.p. $\mathbf{P}$ iff under a suitable order and equivalent norm, G becomes a Banach lattice with order continuous norm and topological unit; $\mathbf{P}$ is then necessarily a complete B.a.p. We remark also that while the norm (*) of G is independent of the choice of the cyclic vector x_0, the order of E is not, in general; however, for any pair of cyclic vectors the corresponding orderings of G are Riesz isomorphic (Rall [14]). Moreover, Rall [14] has shown that τ-completeness of $\mathbf{P}$ suffices to ensure all conclusions of the preceding theorem with the sole exception of the order continuity of the norm.

Let us record a representation theorem for cyclic Banach spaces, whose proof results at once from Theorems 2.3 and II.4.3; of course, an analogous weaker result follows from Theorem 1.3.

PROPOSITION 2.4: *Let G be a cyclic Banach space with respect to a σ-complete B.a.p. $\mathbf{P}$. Then G is isomorphic[†] to a Banach lattice $\hat{G}$ of continuous numerical functions on the Stone space K of $\mathbf{P}$. The lattice $\hat{G}$ possesses order continuous norm and contains $C(K)$ as a dense (lattice) ideal, and each $P \in \mathbf{P}$ acts on G as multiplication by the characteristic function of some open-and-closed subset of K (and conversely).*

Moreover, it is not difficult to show that G is an ideal of the Riesz space $C_\infty(K)$ of all continuous numerical functions on K that assume the values $\pm \infty$ on rare subsets of K only (note that K is Stonian).

3. The Operator Algebra Generated by a B.a.p. In this final section we collect some results on B.a.p.'s into which Theorem 2.3 does not enter explicitly but to which it provides easy access, thus giving an example of the general usefulness of Banach lattice theory. We recall that an abstract Boolean algebra is said to satisfy the *countable chain condition* if every orthogonal subset of A is countable. A σ-complete Boolean algebra A satisfying this condition is complete, and each directed ($\leq$) subset of A contains a cofinal subset which is countable (see, e.g., [16, Chapter II, Exercise 4] for these facts and their bearing on Riesz spaces). We must keep in mind, however, that the definitions of σ-completeness and completeness for B.a.p.'s adopted above (§2) include appropriate continuity requirements.

THEOREM 3.1: *Let $\mathbf{P}$ be a σ-complete B.a.p. on a Banach space G. If there exists a cyclic vector in G or if G is separable, then $\mathbf{P}$ is a complete B.a.p. and satisfies the countable chain condition.*

Proof: In the presence of a cyclic vector $x_0 \in G$, the completeness of $\mathbf{P}$ as a B.a.p. is contained in Theorem 2.3 above. Suppose now that B is an orthogonal subset of $\mathbf{P}$; then $B_0 := \{Px_0 : P \in B\}$ is an orthogonal subset of E (see 2.3) contained in $[0, x_0]$. If B_n ($n \in \mathbf{N}$) denotes the subset of B_0 whose elements have norm $> n^{-1}$, then by Theorem I.4.4 (f), B_n is finite and hence, B is countable.

[†]The algebraic and lattice operations are to be understood as in Theorem II.4.3.

Via the countable chain condition, we can now replace the existence of a cyclic vector by the requirement that G be separable. Let $\{x_n: n \in \mathbf{N}\}$ denote a countable dense (or total) subset of G and let G_n denote the closed linear hull of $\{Px_n: P \in \mathbf{P}\}$. The restrictions of all $P \in \mathbf{P}$ to G_n form a σ-complete B.a.p. $\mathbf{P}_n$ which is obviously isomorphic to the Boolean quotient $\mathbf{P}/\mathbf{J}_n$, where $\mathbf{J}_n$ denotes the Boolean ideal $\{P \in \mathbf{P}: Px_n = 0\}$. By the Boolean homomorphisms $q_n: \mathbf{P} \to \mathbf{P}/\mathbf{J}_n$ we can construct a Boolean homomorphism $q: \mathbf{P} \to \Pi_n \mathbf{P}_n$, letting $q(P) := (q_1(P), q_2(P), \ldots)$. Clearly q is injective, thus $\mathbf{P}$ is isomorphic to a Boolean subalgebra of $\Pi_n \mathbf{P}_n$. By the first part of the proof, each $\mathbf{P}_n$ satisfies the countable chain condition and hence, so does $\mathbf{P}$. (If B, $0 \notin B$, is an orthogonal subset of $\mathbf{P}$, let $B_1 := \{P \in B; q_1(P) \neq 0\}$, $B_2 := \{P \in B \setminus B_1: q_2(P) \neq 0\}$, and so on; then each of the sets B_n ($n \in \mathbf{N}$) must be countable, and $B = \bigcup_n B_n$.) Since $\mathbf{P}$ is σ-complete by hypothesis, $\mathbf{P}$ is complete as an abstract Boolean algebra. Moreover, if (P_α) is a directed ($\leq$) family in $\mathbf{P}$ and if $P = \sup_\alpha P_\alpha$, then to see that $\lim_\alpha P_\alpha x = Px$ for each $x \in G$, it suffices to consider the cyclic subspace generated by x and $\mathbf{P}$, and to apply Theorem 2.3. □

PROPOSITION 3.2: *Let G be a cyclic Banach space with respect to the σ-complete B.a.p. $\mathbf{P}$. A closed vector subspace H of G is $\mathbf{P}$-invariant (i.e., invariant under each $P \in \mathbf{P}$) if and only if $H = PG$ for some $P \in \mathbf{P}$. Moreover, the algebraic sum of any finite number of closed $\mathbf{P}$-invariant subspaces is closed in G.*

Proof: Since $\mathbf{P}$ is commutative, it is clear that each subspace $H = PG$ is $\mathbf{P}$-invariant; moreover H is closed, since $H = (1 - P)^{-1}(0)$. To prove the converse, we employ the representation (2.4). It will be enough to show that the subspace $\hat{H}$ corresponding to H under $G \to \hat{G}$ is an ideal of the Banach lattice $\hat{G}$; for by Theorem 2.3 and Proposition I.4.5, every closed ideal $\hat{H}$ is a projection band of $\hat{G}$ and hence of the form $\hat{H} = P\hat{G}$ for some $P \in \mathbf{P}$. Let $f \in \hat{H}, g \in \hat{G}$ satisfy $|g| \leq |f|$; we must show that $g \in \hat{H}$. Given $\epsilon > 0$, there exist a partition $\{U_i: i = 1, \ldots, n\}$ of K into open-and-closed sets and real numbers α_i, β_i such that $\|\Sigma \alpha_i \chi_i - f\| < \epsilon, \|\Sigma \beta_i \chi_i - g\| < \epsilon$, where χ_i is the characteristic function of U_i. (The Riesz space S of continuous functions with finite range is dense in $C(K)$ (cf. proof of 1.1) and, a fortiori, in $\hat{G}$.) Because of $|g| \leq |f|$, it can be arranged that $|\beta_i| \leq |\alpha_i|$ for $i = 1, \ldots, n$. Denoting by $P_i \in \mathbf{P}$ the projection defined by multiplication with χ_i, letting $\gamma_i := \beta_i/\alpha_i$ if $\alpha_i \neq 0$ and $\gamma_i := 0$ otherwise, and putting $T := \Sigma \gamma_i P_i$, we obtain $\Sigma \beta_i \chi_i = T(\Sigma \alpha_i \chi_i)$. Since $|\gamma_i| \leq 1$ for all i and since the norm of $\hat{G}$ is a Riesz norm, it follows that T has operator norm ≤ 1. Evidently $T\hat{H} \subset \hat{H}$, and so from $\inf_{h \in \hat{H}} \|\Sigma \alpha_i \chi_i - h\| < \epsilon$ it follows that $\inf_{h \in \hat{H}} \|\Sigma \beta_i \chi_i - h\| < \epsilon$. Thus $\inf_{h \in \hat{H}} \|g - h\| < 2\epsilon$ for all $\epsilon > 0$, and this implies that $g \in \hat{H}$.

We have now shown that the set of all **P**-invariant closed subspaces of G corresponds bijectively to the set of all projection bands of $\hat{G}$; thus the second assertion follows at once from Proposition I.2.3. □

Our final objective is a characterization of $A(\mathbf{P})$ (see 1.1) as a subalgebra of $\mathcal{L}(G)$, G being supposed cyclic with respect to a σ-complete B.a.p.; this characterization will again be based on Proposition 2.4. It is necessary to realize that $C_\infty(K)$ (where K is the extremally disconnected Stone space of **P**; see the remark after 2.4) is an algebra under pointwise multiplication of functions on sets of finiteness; this follows at once from the fact that each $f \in C_\infty(K)$ is finite on a dense open subset of K, and from a slight modification of Proposition II.1.6 (bounded closed intervals of **R** being replaced by the two-point compactification $\overline{\mathbf{R}}$). Now as we have observed, $\hat{G}$ is a lattice ideal of $C_\infty(K)$; but it is not necessarily a subalgebra (and much less an algebraic ideal). Still, $\hat{G}$ being a lattice ideal of $C_\infty(K)$, the multiplication of functions is possible within $\hat{G}$ if at least one factor is in $C(K)$. For the announced characterization, we need a converse; as before, S denotes the set of functions with finite range in $C(K)$.

LEMMA 3.3: *Suppose $g \in \hat{G}$ is a function such that $gf \in \hat{G}$ for all $f \in S$ and such that $\sup\{\|gf\| : f \in S, \|f\| \leq 1\} < +\infty$. Then g is finite (or equivalently, $g \in C(K)$).*

Proof: Suppose that g is not finite throughout K; then we can assume (by a change of sign, if necessary) that all of the sets $A_k := \{s \in K : g(s) > k\}$ ($k \in \mathbf{N}$) are non-void. Since K is Stonian (**P** being complete by 3.1), $\bar{A}_k$ is open-and-closed; denote by χ_k the characteristic function of $\bar{A}_k$ and define $h_k := \chi_k / \|\chi_k\|$. Then we have $gh_k \geqslant kh_k$ and, since the norm of $\hat{G}$ is a Riesz norm, we obtain $\|gh_k\| \geqslant k$ for all $k \in \mathbf{N}$, a contradiction. □

THEOREM 3.4 (Bade): *Let G be a cyclic Banach space with respect to a σ-complete B.a.p.* **P**. *For an operator $T \in \mathcal{L}(G)$, the following assertions are equivalent*:

(a) *T belongs to the norm closed operator algebra $A(\mathbf{P}) \subset \mathcal{L}(G)$ generated by* **P**.

(b) *T belongs to the strongly closed operator algebra $W(\mathbf{P}) \subset \mathcal{L}(G)$ generated by* **P**.

(c) *T leaves invariant every* **P**-*invariant closed subspace of G*.

Proof: The implications (a)⇒(b)⇒(c) are trivial. (c)⇒(a): Suppose T satisfies (c); we consider T as a bounded linear operator on the Banach lattice $\hat{G}$ (see 2.4), recalling that each $P \in \mathbf{P}$ acts by multiplication with some χ_U ($U \in \mathbf{Q}(K)$). By Proposition 3.2, (c) means that $TP = PT$ for each $P \in \mathbf{P}$. Thus if e denotes the constant-one function on K, we obtain $T\chi_U = T(\chi_U e) = (T\chi_U)e = \chi_U(Te)$. Thus $Tf = f(Te)$ for each $f \in S$; since T is bounded, Lemma 3.3 implies that $Te \in C(K)$. Moreover, Te is the limit (in the supremum norm of $C(K)$) of a sequence (g_n) in S; clearly, each of the multiplication operators $f \mapsto fg_n$ is in $A(\mathbf{P})$. Thus by Proposition 1.1 we obtain $T \in A(\mathbf{P})$. □

COROLLARY: *On the representation $\hat{G}$ of G* (2.4), *the isomorphism $A(\mathbf{P}) \cong C(K)$ of* (1.1) *is realized by considering each $T \in A(\mathbf{P})$ as multiplication with the corresponding function of $C(K)$*.

Theorem 3.4 is due to Bade [1] (see also [5]) and is valid, without the assumption that G be cyclic, for any complete B.a.p. **P**. As shown by Rall [14], this more general version can be obtained, with the present means and little additional effort, for a τ-complete **P** which is complete as an abstract Boolean algebra. He also gave necessary and sufficient conditions for an analogue of Theorem 2.3 to hold without assuming the existence of a cyclic vector in G.

REFERENCES

1. W. G. Bade, "On Boolean algebras of projections and algebras of operators," *Trans. Amer. Math. Soc.*, **80** (1955), 345-360.
2. R. G. Bartle, *The Elements of Integration*, Wiley, New York, 1966.
3. Garrett Birkhoff, *Lattice Theory*, Amer. Math. Soc. Colloq. Publ., 25, 3rd ed., Providence, 1967.
4. D. I. Cartwright, "Extension of positive operators between Banach lattices," *Mem. Amer. Math. Soc.*, **164** (1975).
5. N. Dunford and J. T. Schwartz, *Linear Operators, Part III: Spectral Operators*, Wiley, New York, 1971.
6. D. Fremlin, "Tensor products of Archimedean vector lattices," *Amer. J. Math.*, **94** (1972), 777-798.
7. _____, "Tensor products of Banach lattices," *Math. Ann.*, **211** (1974), 87-106.
8. A. Grothendieck, "Une caractérisation vectorielle-métrique des espaces L^1," *Canad. J. Math.*, **7** (1955), 552-561.
9. A. Goullet de Rugy, "La structure idéale des M-espaces," *J. Math. Pures Appl.*, **9** (51) (1972), 331-373.
10. H. P. Lotz, "Extensions and liftings of positive linear maps on Banach lattices," *Trans. Amer. Math. Soc.*, **211** (1975), 85-100.
11. W. A. J. Luxemburg and A. C. Zaanen, "Notes on Banach function spaces," 1-13, *Indag. Math.*, **25** (1963), **26** (1964).
12. _____, *Riesz Spaces I*, North-Holland, Amsterdam, 1971.
13. _____, "The linear modulus of an order bounded linear transformation," *Indag. Math.*, **33** (1971), 422-447.
14. C. Rall, "Über Boolesche Algebren von Projektionen", *Math. Z.*, **153** (1977), 199-217.
15. H. H. Schaefer, *Topological Vector Spaces*, Springer, Berlin-Heidelberg-New York, 1971.
16. _____, *Banach Lattices and Positive Operators*, Springer, Berlin-Heidelberg-New York, 1974.
17. Z. Semadeni, *Banach Spaces of Continuous Functions*, Polish Scient. Publ., Warsaw, 1971.
18. A. I. Veksler, "Cyclic Banach spaces and Banach lattices," (English Trans.), *Soviet Math. Dokl.*, **14** (1973), 6, 1773-1779.

INDEX

Absolute projection constant, 56
Abstract Lebesgue space, 184
Abstract M-space, 179
Affine combination, 142
   independence, 141
   map, 119
AL-space, 184
Algebra, Boolean, 209
Algebra numerical range, 15, 35
Algebra, uniform, 97
AM-space, 179
Approximate point spectrum, 4
Approximation problem, 82
   property, 82, 98
Archimedean ordering, 160
Ascent, 20

B^*-algebra, 39
Banach lattice, 162
   limit, 14
Band, projection, 165
Barycenter, 118
Base, for a cone, 144
Basic sequence, 89
Basis, in a Banach space, 90
   constant, 90
Berman-Marcinkiewicz equation, 61
Bernstein operators, 53
Best approximation, 50
Bimorphism, 198
Biorthogonal, 89
Bochner integrable function, 196
Boolean algebra, 209
Borel sets, 120
Bounded approximation property, 84
   compact approximation property, 98

C^*-algebra, 39
Calkin algebra, 45
Cantor set, 88

Carrier, of an operator, 68
Chain condition, 217
Character, of a group, 102
Compact approximation problem, 82, 98
   Hermitian operator, 33
   operator, 82
Complemented ideal, 165
   subspace, 56, 83
Complete Boolean algebra, 213
Completely monotone function, 155
Concave function, 122
Cone, 142, 144, 161
Convex combination, 116
   function, 122
   hull, 151
   set, 115
Corner, of a convex set, 5
Countable chain condition, 217
Cross-norm, 204
Cyclic Banach space, 210
   vector, 210

Decomposition, finite dimensional, 94
Dedekind complete, 163
Dirichlet kernel, 61
Disc algebra, 25
Discrete probability measure, 118
Doubly stochastic matrix, 136

Eigenvalue, 4
Envelope, 122
Equivalent bases, 91
Ergodic mapping, 133
   theory, 132
Essential numerical range, 45
   spectrum, 45
Extension property, 194
Extremal algebra, 42
Extremally disconnected, 59, 180
Extreme point, 6, 116

Finite dimensional decomposition, 94
  rank operator, 82
Fredholm operator, 45
Freudenthal unit, 182

Gershgorin discs, 22

Haar subspace, 64
Hermitian operator, 20
Homomorphism, metric, 194
Homothetic image, 147
Hyperstonian space, 188
Hyponormal operator, 6

Ideal, in a Riesz space, 163
Injective object, 198, 206
Invariant measure, 133
Isometry, 14
Isomorphic Banach spaces, 83
Isotone operator, 174

Knots, 66
Kreĭn-Mil'man theorem, 68

Lagrange interpolation, 55
Lattice ideal, 161
  order, 144
Least squares, 55
Lebesgue constant, 62
  function, 65, 75
l_p-sum of Banach spaces, 96

M-space, abstract, 179
Mean trace, 101
Measure-preserving transformation, 132
Metric homomorphism, 194
Minimal projection, 60
Modulus, 161

Norm, order continuous, 171
  p-additive, 182
  Riesz, 162
Normal operator, 2
Normalized state, 34
Numerical function, 192
  radius, 2, 41

Numerical range, 2
  algebra, 15, 35
  essential, 45
  spatial, 11

Operator, Bernstein, 53
  compact, 82
  finite rank, 82
  Fredholm, 45
  Hermitian, 20
  hyponormal, 6
  normal, 2
  order bounded, 175
  positive, 2, 174
  projection, 54, 83
  self adjoint, 2
  unitary, 2
Order bounded operator, 175
  bounded linear form, 167
  bounded set, 162
  complete, 163
  continuous form, 168
  continuous norm, 167
  dual, 167
  lattice, 144
  unit, 182
Ordered vector space, 160
Orthogonal elements, 161
  projection, 56, 72
Orthogonality, 21

p-additive norm, 182
Partial order, 143
π-norm, 195
π-tensor product, 195
Point mass, 117
  spectrum, 4
Positive cone, 161
  operator, 2, 174
Power inequality, 9
Printer, 15
Probability measure, 117, 120
Projection, 54, 83
  band, 165
  constant, 56
  minimal, 60

INDEX

Projective cone, 204
Proximity map, 51

Quasi-interior point, 189
Quasi-Stonian space, 180

Rademacher functions, 62, 103
Regular bilinear form, 203
    measure, 120
    operator, 176
Relative projection constant, 56
Relatively uniformly complete, 200
Resultant, 118
    map, 141
Riesz bimorphism, 198
    homomorphism, 163
    norm, 162
    representation theorem, 120
    space, 160
r-norm, 176

Self-adjoint operator, 2
Semi-continuous, 122
Semi-inner product, 11
Separation theorem, 138
σ-complete Boolean algebra, 213
Simplex, 142, 146
Smooth point, 36
Spatial numerical range, 11
Spectral radius, 5
Spectrum, 4
    essential, 45

Spline function, 66
State, normalized, 34
Stochastic matrix, 24, 136
Stone space, of a Boolean algebra, 209
Stonian space, 180
Strictly convex function, 122
Structure space, 190
Symmetric cross-norm, 204

τ-complete Boolean algebra, 213
Topological unit, 189
Trace, mean, 101

Uniform algebra, 97
Uniformly complete, relatively, 200
Unit, topological, 189
Unital Banach algebra, 34
Unitary operator, 2
Upper envelope, 122
Upper semi-continuous, 122

Valuation, 191
Vector lattice, 160
Vertex, 36

Walsh function, 102
Weak convergence, 127, 129
    order unit, 182
Weak-* operator topology, 56
Weak-* topology, 86, 127
Width of a set, 51

INDEX OF SYMBOLS

$A(K)$, 120
a.p., 82
$\text{apSp}(T)$, 4

b.a.p., 84
B.a.p., 209
b.c.a.p., 98
BL, BS, 198
$\mathbf{B}(E)$, 165
$BL(X)$, 10
$\mathscr{B}(G,H)$, 195
$\mathscr{B}^r(E,F)$, 203

c.a.p., 82
$\text{co } A, \overline{\text{co}} A$, 151

$D(I)$, 16
$D(x)$, 11
$\delta_n(K)$, 51
Δ, 88

$\text{Ea}(K)$, 42
$E(K,Y)$, 51
E_{oc}^*, 168
$\text{ext } K$, 71, 116
$\text{ext}(K)$, 6
ε_x, 117

f.d.d., 94

$\text{Her}(A)$, 37

$\mathbf{I}(E)$, 164

$KL(X)$, 45
$l_\infty(S)$, 58
$L^r(E,F)$, 175

$\mathcal{L}^r(E,F)$, 117
$\lambda(Y), \lambda(Y,X)$, 56
Λ_P, 65, 75

$\text{pSp}(T)$, 4
P_1, 130
$\tilde{P}_1$, 134

$\mathbf{Q}(X)$, 209

$r(T)$, 5
ρ_n, 61, 62

$S(X)$, 10
$\text{Sp}(T)$, 4
$\text{Sp}_e(T)$, 45

$\hat{t}$, 55

U, 85

$v(a)$, 41
$V(a)$, 35
$V(T)$, 11
$V_e(T)$, 45
$V(B,T)$, 10

$w(T)$, 2
$W(G)$, 102
$W(T)$, 1, 2

$\perp$, 21, 161
$\sim$, 96
$\overline{\otimes}$, 200
$\tilde{\otimes}_p$, 204
$\otimes_\pi, \tilde{\otimes}_\pi$, 195